DES
VERSIONS

KC-135 STRATOTANKER

in detail & scale

A DETAIL & SCALE AVIATION PUBLICATION

Alwyn T. Lloyd

KALMBACH K BOOKS

Airlife Publishing Ltd.
England

CONTRIBUTORS AND SOURCES:

Robert D. Archer
CPT Robert Barr, USAF
John T. Bohn
Harold V. Brackin
Peter M. Bowers
Dr. Dennis F. Casey
LTC R. Chubbick, USAF
1LT Kim Cloe, USAF
Thomas R. Cole
COL P. Wayne Corbett, USAF (Ret.)
MAJ R. Cunningham, USAF
MSGT Alan R. Doty, USAF
Jim Dunham
COL Michael D. Edwards, USAF (Ret.)
Norman R. Filer
Rob Fontaine
John F. Fuller
Harry S. Gann
Richard Geist
William Grossclose
John B. Grasso
COL Thomas J. Harris, USAF
CPT S.M. Hicky, USAF
J.C. Hopkins
Earl H. Jacobson
Lloyd S. Jones
Bert Kinzey
Arthur B. Kline
Walter Koons
Richard Lane
Ray Leader
William H. Lloyd
LTC Sandy Mackelprang, Arizona ANG (Ret.)
Clark Macomber
D. Bruce McCullough
David W. Menard
A. "Monty" Montemayor
Ned Nelson
Marilyn A. Phipps
Dennis R. Pitts
Otto Praeger
Mike Quan
CPT Thomas J. Ramsey, USAF
Doug Remington
BG Michael G. Routsala, USAF
Victor D. Seely
James Sheperd
COL William R.Sherer, Arizona ANG
Robert L. Snellenberg
CPT H. Sunderland, USAF
Norman E. Taylor
COL Wayne Wachsmuth, USAF (Ret.)
COL Arnold L. Weinman, USAF
Gordon S. Williams
CPT David Willis, USAF
David H. Wilson
Len Wright

U. S. Air Force Association
The Boeing Company - Public Relations & History Offices
Douglas Aircraft
E-Systems
Eighth Air Force - Public Affairs & History Offices
Fifteenth Air Force - Public Affairs & History Offices
Oklahoma City Air Logistics Center (AFMC), Tinker AFB, OK
Strategic Air Command - Public Affairs & History Offices
6th Strategic Reconnaisance Wing (SAC), Eielson AFB (SAC), AK
9th Strategic Reconnaissance Wing (SAC), Beale AFB, CA
22nd Air Refueling Wing (SAC), March AFB, CA
92nd Bombardment Wing (SAC), Fairchild AFB, WA
93rd Bombardment Wing (SAC), Castle AFB, CA
112th ARefG, Pennsylvania ANG
128th ARefG, Wisconsin ANG
157th ARefG, New Hampshire ANG
160th ARefG, Ohio ANG
161st ARefG, Arizona ANG
171st ARefW, Pennsylvania ANG
186th ARefG, Mississippi ANG
301st ARefW, Malmstrom AFB (SAC), MT
320th Bombardment Wing (SAC), Mather AFB, CA
453rd ARefG, Fairchild AFB (AMC), WA
940th ARefG (AFRes), Mather AFB, CA

Most photographs in this publication are credited to their contributors. Photographs with no credit indicated were taken by the author.

FIRST PRINTING

Published in United States by
Kalmbach Publishing Co.
21027 Crossroads Circle
P.O. Box 1612
Waukesha, Wisconsin, 53187

Library of Congress Cataloging-in-Publication Data

Lloyd, Alwyn T.
KC-135 Stratotanker : in detail & scale / Alwyn T. Lloyd.
p. cm. -- (D & S ; vol. 44)
"A Detail & scale aviation publication."
"Also includes C-135 cargo versions."
Includes bibliographical references.
ISBN 0-89024-203-8
1. KC-135 (Tanker aircraft) I. Title. II. Series.
UG1242.T36L58 1994
623.7'46--dc20 94-11660

Published in Great Britain
by Airlife Publishing Ltd.
7 St. John's Hill, Shrewsbury, SY1 1JE

British Library Cataloging in Publication Data

A catalog record for this book
is available from the British Library.

ISBN 1-85310-641-0

Front cover:** KC-135A-24-BN, 62-3564, was the 615th aircraft off the assembly line and was delivered to Lockbourne AFB, Ohio, on 21 June, 1963. It was painted in the overall Corogard silver finish with Aircraft Gray engine cowls. Note the boom latch and safety cable and the colored stripes on the telescoping portion of the boom. This aircraft was converted to a KC-135R in December 1985.* ***(Boeing-Wichita TW1075-238C)

***Back cover:** KC-135A-20-BN, 61-0300, was delivered to Offutt AFB, Nebraska, on 8 June, 1962. When this photograph was taken, the aircraft was serving as the flagship for the Fifteenth Air Force at March AFB, California. The green and white tail band was used by the 22nd Air Refueling Wing. Note the white boom which was peculiar to this aircraft. This KC-135A was converted into a KC-135R in April 1990.*

INTRODUCTION

This short-tailed KC-135A, 58-0128, reveals a natural metal finish with Aircraft Gray applied to the wing center section, engine cowls, nose gear doors, and much of the main gear doors. Da-glo orange paint was applied to the nose, wing tips, and aft fuselage band. The yellow refueling alignment stripe and director lights are also visible in this view.* *(Boeing via G.S. Williams)

The Boeing KC-135A Stratotanker is truly a plane of a thousand faces. While it was manufactured in seven series, no less than forty-two variants have been produced through Air Force-directed modifications. To study the progression one needs a detailed road map in order to have any degree of certainty in following the trail.

It was this airplane which placed the Strategic Air Command (SAC) in the truly all-jet arena. The KC-135 provided global reach for America's heavy bombardment forces, and it made trans-ocean deployments for all U. S. fighter units a practical reality. It gave fully loaded Military Airlift Command (MAC), and now Air Mobility Command (AMC), transports the long legs required to support U. S. military units anywhere in the world. This was exemplified by its phenomenal performance in the deployment, then the combat operations, and finally the return of American forces during Operations Desert Shield, Desert Storm, and Proud Return. The KC-135 is truly a national asset.

Because of the prolific progeny of the original KC-135A Stratotanker, there will be two volumes in the Detail & Scale Series to provide reasonable coverage of the many variants of this remarkable aircraft. This first volume briefly covers the development of the art and science of aerial refueling, then provides a basic description of the KC-135 systems and structures. It also takes a look at each of the tanker and transport variants of the aircraft. The RC-135A is also included here because of its unique mission, its subsequent use as a transport, and finally its conversion to a tanker. Specifically, this volume covers the following aircraft: KC-135A, C-135F/C-135FR, C-135A/C-135E, C-135B/C-135C, RC-135A, KC-135D, KC-135Q, KC-135E, KC-135R, and the VC-135B. The differences between the KC-135 and the Boeing Model 707, as described in Detail & Scale Volume 23, is also reinforced.

As with all volumes in the Detail & Scale Series, the focus of this book is on the many physical details of the aircraft. Never before has a publication provided such extensive coverage of the Stratotanker's details. From the cockpit to the refueling boom, from the top of the tail to the landing gear, and from wingtip to wingtip, all of the important features of the aircraft are described and profusely illustrated with numerous photographs and drawings. Modifications to the airframe during its operational life and the differences between variants are also thoroughly covered.

From a modeler's perspective, building a model of such a large aircraft can require a great deal of time and effort. With no armament to "dress up" the model, the choices of an accurate paint scheme and attractive markings are important in order to insure the desired eye appeal when the model is completed. Therefore, this volume also devotes several pages to the basic color schemes and markings as they have evolved over the years.

The modeler's section includes reviews of all of the kits that are currently available to the serious scale modeler. Of particular interest is a look at the recently released 1/72nd scale kits of the Stratotanker from ERTL. Conversion parts and after-market decals are also included in the modeler's section.

A second volume will provide coverage of the special-purpose variants of the aircraft. Included will be the EC-135s, RC-135s, JKC-135s and NKC-135s; these being the airborne command posts, reconnaissance, and the temporary and permanent test aircraft, respectively.

KC-135 HISTORY

This KC-135A was part of the Boeing test fleet, and it carries a flight test probe. It was photographed lifting off from Boeing Field, Seattle, Washington, and the landing gear can be seen in the retraction process. (Boeing P18559)

The art of aerial refueling had its genesis during the barnstorming days when two aircraft would fly in formation, and one would drop a rope ladder to the other. An intrepid individual would climb from one ship to the other with a five-gallon can of gasoline strapped to his back. However, this was hardly more than an air show stunt with little practical value.

The most famous aerial refueling of the pioneering days occurred when the U.S. Army Air Corps had an American Aircraft Corporation Fokker C-2, named the **Question Mark**, stay aloft for over 140 hours. During this flight in January 1929, Major Carl A. "Tooey" Spaatz, Captain Ira C. Eaker, and three other Air Corps aviators made forty-three refueling contacts over southern California.

First USAF Tankers

During the mid-1940s, the U. S. Air Force attempted aerial refueling in earnest when it contracted with Boeing to adapt the hose system developed by British Flight Refueling, Ltd. for use on the B-29. These hose tankers were designated KB-29Ms. This system required that the receiver fly above the tanker trailing a cable with a grapnel hook while the tanker extended a cable lead with a ball end in what was known as a cross-over maneuver. After the grapnel hook snared the tanker's cable, the two aircraft would change altitudes so that the tanker was above the receiver. Meanwhile the receiver would reel in the tanker's cable which was attached to the refueling hose. The hose was connected in the tail of the receiver, and refueling commenced. The most famous of these refuelings was when the **Lucky Lady II**, a Boeing B-50A, made the first non-stop, around-the-world flight in 1949. However, this tedious system did not prove to be very efficient.

The Flying Boom

Boeing then developed the flying boom concept in which a telescoping boom would be installed in the belly of a B-29. A pair of controllable ruddevators were mounted at the aft end of the fixed portion of the boom so that the boom operator could literally fly the boom into the receptacle on the receiver aircraft. With this system, the boom operator controls the hookup while the tanker's pilots control the fuel transfer. A total of 116 Superfortresses were converted to KB-29Ps with this system. The flying boom would later be installed on Boeing KC-97 Stratofreighters. (See Detail & Scale Volume 25 on B-29 derivatives of the B-29.)

Operations with the KC-97 were most efficient when refueling piston-powered B-29 and B-50 bombers. However, with the advent of the jet bomber force of Boeing B-47 Stratojets and B-52 Stratofortresses, refueling again became problematic. Basically, the aircraft had two different operating envelopes. There were speed differences to contend with, because the jets were 600-mph class aircraft, whereas the tankers were in the 300-mph class. The faster jets tended to stall when refueling behind the tankers. The bombers also worked efficiently at higher altitudes that the tankers could not achieve. Therefore, the bombers had to descend to lower altitudes to refuel. An interim solution was to fly a toboggan maneuver in which the tankers would accelerate in a shallow dive as they lost weight. At the same time, the bombers could fly faster as their weight increased with the transferred fuel. This was important to prevent stalling, because the bomber required greater airspeed as its weight increased. This type of flying was sporty at best. There were horror stories of bombers underrunning the tankers and shearing off booms. Sometimes, parts of the bombers would return to base within the tankers and vice-versa. Certainly, a better solution was needed.

Recognizing the problem, Boeing, in a private venture, began design studies based on the Model 367 (KC-97) to resolve the problems of operational incompatibility. The result was the Model 367-80, a swept-wing, four-engined jet aircraft. The 367-80 had the same body diameter as earlier series airplanes. While this made a good test bed, it was impractical for operational use. The Model 367-80 first flew on 15 June, 1954, and was usually referred to as the Dash 80. Boeing demonstrated the aircraft to both the U. S. Air Force and the airlines. So impressed with its potential, the Air Force ordered twenty-nine tankers which were based on the design of Dash 80. The new tanker was designated the KC-135. This order was placed even before the final design had been established or the Dash 80 aerial refueling capabilities had been demonstrated.

The KC-97 boom system was adapted for installation on the 367-80, and refueling tests were conducted with a B-52. To meet Air Force requirements, the fuselage diameter was increased from 132 to 144 inches. This new airplane was designated the Model 717 by Boeing. A total of 732 KC-135A Stratotankers were produced.

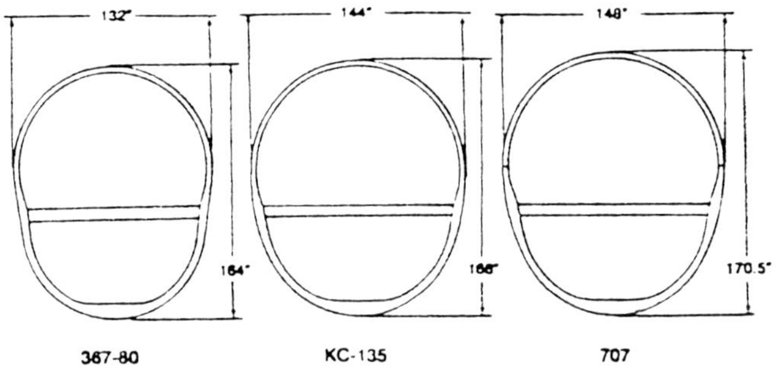

The body diameters were different for the KC-97 (Model 367/Model 367-80), KC-135 (Model 717), and Model 707.

Model 707 versus Model 717

It should not be thought that the KC-135 (Model 717) is a military version of the commercial Model 707. (See Detail & Scale Volume 23 on the Boeing 707 and AWACS.) As stated above, the body diameters are different, and the fuselage cross sections are also different. The Model 717 has an oval cross section, and the Model 707's fuselage has a double lobed cross section. The body length of the Model 717 is 128' 10," while the 707-100/-200 length is 138' 10."

Immediately obvious are the cabin windows on the Model 707. No cabin windows were installed on any version of the -135 with the exception of those in the emergency exit hatches located at the overwing area on both sides and in the aft right waist. All -135s have the boom pod bulge, a crew entry hatch beneath the cockpit, and a large cargo door in the left side of the forward fuselage.

The KC-135s are powered by turbojet engines which utilize direct engine bleed air for cabin pressurization. Therefore, on the KC-135As, the engine struts fair directly down onto the cowls at the leading edge of the strut. Both turbojet and turbofan-powered -135s have been produced. Early Model 707s were turbojet-powered; whereas all later 707s were powered by turbofan engines. The 707s with turbojets utilized engine bleed-driven turbocompressors for cabin pressurization and are recognizable by the forward extension or hump on the strut fairings which have a small opening above the engine inlet cowl. Fan-powered -135s have electrical generators in these humped struts.

KC-135As were delivered with Pratt & Whitney J57 turbojet engines. Similar engines, designated JT3Cs, which were equipped with sound suppressors, were installed on early Model 707 commercial jetliners. A number of KC-135Bs, RC-135Bs, and C-135Bs were delivered to the Air Force with Pratt & Whitney TF33 turbofans. The commercial equivalent to the TF33 was the JT3D, however, these engines had one more compressor stage. Relative sea level static thrust for these engines is:

J57	13,750 pounds
JT3C	12,500 pounds
TF33	18,000 pounds
JT3D	18,000 pounds

The wing planforms of the Model 717 and Model 707-100 are virtually identical, and have a span of 130' 10." The wing root chords are 338," while the tip chords are 112." The wing leading and trailing edges are straight, except for the trailing edge fillet.

In the 1950s, Boeing and the aircraft industry were trying to gain an understanding of new flight envelopes and materials. To achieve a greater strength-to-weight ratio, the Model 717 structure was designed under what was known as a safe-life philosophy. In contrast, the subsequent Model 707 structures were built to a fail-safe philosophy as required by the Federal Aviation Administration. Taking into account the aircraft utilization, such an approach is not imprudent. A military aircraft is expected to carry high payloads great distances, but is not flown as many hours per year as a commercial airliner. Roughly, a military aircraft is flown ten percent of the hours an airliner is used.

There were five versions of the Model 717 delivered to the U. S. Air Force. Next came the two Model 739s which were employed in the reconnaissance roles. From these seven production series of airplanes sprang an almost unbelievable number of derivations whose differences vary from subtle to obtuse.

Production Series of the -135 Family

By far, the variant of this design which was produced in the greatest numbers was the KC-135A, with 732 examples being built. The basic airframe, including the sizes and shapes of the fuselage and wings, remained constant throughout the production run for all versions of the aircraft, whether they were Model 717s or 739s.

A total of 820 -135 were produced by Boeing in seven series. The as-produced series were:

Number Built	USAF Designation	Boeing Model Number	Customer
732	KC-135A	717-100	SAC
		717-146	
		717-148	
17	KC-135B	717-166	SAC
10	RC-135B	739-445B	SAC
15	C-135A	717-157	MATS
30	C-135B	717-158	MATS
4	RC-135A	739-700	MATS
12	C-135F	717-164	France
820 Total			

The Air Force made its procurement intentions for the KC-135 known on 5 August, 1954, and the contract was signed on 5 October of the same year. The first airplane rolled out of the factory on 18 July, 1956, and made its maiden flight on 31 August, 1956. So versatile was the airframe that the Air Force ordered numerous conversions so that no less than forty series may be identified by prefix and suffix designations.

The first KC-135A was delivered to the Air Force on 30 April, 1957, where it became part of the test fleet. The first KC-135A Stratotankers entered service with the 93rd Air Refueling Squadron (ARefS) of the 93rd Bombardment Wing (BMW) at Castle AFB, California. The first aircraft, 57-3157, arrived for duty on 28 June, 1957. Throughout the year more tankers found their way into service with the 42nd ARefS of the 42nd BMW at Loring AFB, Maine, and the 99th ARefS of the 99th BMW at Westover AFB, Massachusetts. The last of the 732 production KC-135As, 64-14840, was delivered to the 380th ARefS of 380th BMW at Plattsburgh AFB, New York, on 12 January, 1965. Another twelve C-135Fs were produced for France's Armee de l' Air. These aircraft were essentially identical to the KC-135A except for the French markings. Next in the series were the C-135As and C-135Bs that were produced for the Military Air Transport Service (MATS).

Another variant that was produced for SAC was the turbofan-powered RC-135B. Aircraft in this series went immediately to the Martin Aircraft Company in Baltimore, Maryland, where they were converted to RC-135Cs with large side-looking aerial radar (SLAR) cheeks.

Last in the series were four RC-135As for MATS. Delivery was delayed for several years while the camera system was developed and tested. These aircraft served with the 1370th Photo Mapping Wing, Air Photographic and Charting Service, and they were based at Turner AFB, Georgia.

SAC Record Flights

The Strategic Air Command, under the leadership of General Curtis E. LeMay, immediately set out to break former records and establish new ones with the KC-135. On 11-12 November, 1957, General LeMay, U. S. Air Force Vice-Chief of Staff, flew KC-135A, 55-3126, and established an official world record nonstop, unrefueled flight of 6,322.85 miles. The flight began at Westover AFB and concluded at Buenos Aires, Argentina. The total flying time was thirteen hours, two minutes and fifty-one seconds. An extra 1,000 miles was added to the normal flight path by flying around the hump of Brazil. For this feat, General LeMay was awarded the Harmon International Trophy. On the return trip, flown on 13 November, General LeMay flew a more direct route of 5,204 miles, and established a world course speed record from Buenos Aires to Washington, D.C. These flights were in conjunction with Operation Long Legs, the U. S. Air Force's participation in Argentina's annual Aeronautics Week.

In Operation Top Sail, a pair of KC-135s from the 99th ARefS broke existing speed records in flights from New York to London and back on 27 and 29 June, 1958. Major Burl B. Davenport commanded aircraft 56-3630 on the trip which took five hours, twenty-nine minutes and fourteen seconds. The second aircraft made the trip in five hours, fifty-three minutes, and twelve seconds. A third aircraft crashed on takeoff from Westover AFB, killing all fifteen aboard, including Brigadier General Donald W. Saunders, 57th Air Division Commander; and LTC. George M. Broutsas, 99th ARefS Commander.

On 24 September, 1958, another 99th ARefS KC-135A, piloted by Captain William H. Howell, lifted a 78,089.5 pound payload to an altitude of 2,000 meters (approximately 1.25 miles) breaking a Soviet Tu-104 record of 44,214 pounds which had been set a few weeks earlier on 6 September.

Operation Jet Stream was flown on 7-8 April, 1958, when KC-135A, 56-3601, from the 93rd ARefS established four world records. The aircraft was under the command of Brigadier General William E. Eubank, Jr., 93rd BMW Commander. The aircraft broke the nonstop, unrefueled, straight line flight record by covering the 10,229.3 miles from Tokyo to Lajes, Azores, at an average speed of 492.262 miles-per-hour. The Tokyo to Washington, D.C. segment was flown in thirteen hours, forty-five minutes, and forty-six seconds.

On 17 September, 1957, Captain Charles E. Gibbs from the 92nd ARefS of the 92nd BMW, at Fairchild AFB, Washington, established four world records. These included a distance in a closed circuit without refueling of 3,125.56 statute miles, a speed record for 2,000 kilometers of 589.278 miles-per-hour, flying a closed circuit with 2,204.6, 4,409.2, 11,023, and 22,046 pound payloads, and a speed record of 587.136 miles-per-hour for 5,000 kilometers on closed circuit with 2,204.6 4,409.2, 11,023, and 22,046 pound payloads.

Airborne Alert

On 1 July, 1960, SAC began testing the airborne command post at Offutt AFB, Nebraska. One of five specially modified KC-135s from the 34th ARefS of the 55th Strategic Reconnaissance Wing (SRW) was placed on ground alert and periodically flown throughout the remainder of the year to determine its ability to takeoff within fifteen minutes. The aircraft, with a general officer aboard, would serve as an alternate command post in the event that SAC's command post was destroyed.

Known as Looking Glass, the airborne command post was placed in continuous operation on 3 February, 1961. LTG John P. McConnell was the first Airborne Emergency Action Officer aboard this aircraft. SAC's last twenty-four hour airborne alert mission was flown on 24 July, 1990. Today, STRATCOM has a modified ground/air alert, in which they fly once in a twenty-four hour period at random times.

Tanker Management

In November, 1961, Headquarters U. S. Air Force designated SAC as the single manager for all KC-135 air refueling operations. As such, SAC was responsible for the refueling support of all fighters assigned to the Tactical Air Command (TAC) as well as their own bombers.

SAC developed the Tanker Task Force based at its headquarters at Offutt AFB. From there all world-wide tanker operations were directed. Routine mission tanker support was scheduled through this office as were all deployments requiring tanker support. In the event of an

KC-135A, 58-0084, was assigned to the 11th BMW and was participating in the 1959 SAC Bomb Competition when this photograph was taken. Later, the aircraft was converted to a KC-135Q. (USAF)

Also participating in the 1959 Bomb Competition was KC-135A, 58-0035, from the 7th BMW. Note the typical black band on the fin. This aircraft was later converted into a KC-135R. (USAF)

emergency, the Tanker Task Force would locate the nearest tanker to provide support to any aircraft in the world. SAC's tanker fleet provided refueling support for USAF, ANG, AFRes, U. S. Navy, and USMC deployments. In addition, they also provided tanker support for NATO forces.

With the reorganization of the Air Force in 1991/1992, most KC-135 and KC-10 tanker assets, along with the Tanker Task Force, were placed under the new Air Mobility Command which is headquartered at Scott AFB, Illinois.

Engineering and maintenance control for all -135s is maintained by the Oklahoma City Air Logistics Center (OCALC) at Tinker AFB, Oklahoma. While much of the depot-level maintenance is conducted at Tinker, a considerable amount of work is contracted out. For a long time the Hayes International Corporation at Birmingham, Alabama, did contract work on the tankers. Of course, Boeing-Wichita does major overhauls including all of the re-engine programs. Additionally, Programmed Depot Maintenance (PDM) is carried out by PEMCO (formerly Hayes International), and at the Oklahoma and Sacramento Air Logistics Centers. Another big player in the modification business is E-Systems in Greenville, Texas, where many of the RC-135 versions are upgraded and maintained.

Combat Operations

On 9 June, 1964, KC-135As began supporting combat operations in Southeast Asia. Nicknamed the Yankee Team Tanker Task Force, four KC-135As refueled eight F-100s on their way to strike Pathet Lao anti-aircraft emplacements on the Plain of Jars in northern Laos. The tanker unit was renamed the Foreign Legion on 3 September.

The 4252nd SW was activated at Kadena AB, Okinawa on 12 January, 1965, to provide tanker support for tactical fighters engaged in combat over Vietnam. Known as Young Tiger, the first mission was flown on 25 January. In February 1965, the first Arc Light B-52 missions were flown with KC-135 tanker support.

KC-135A, 62-3572, was captured on film during the war in Southeast Asia. Note the addition of the SAC Milky Way band and the mission name Ramrod. This aircraft has since been converted to a KC-135R. (USAF)

During 1967, KC-135As operated out of Kadena AB, Okinawa, U-Tapao AB, Thailand, and other bases in the Western Pacific. They flew over 22,000 sorties while transferring 1.1 million pounds of fuel in support of SAC B-52s and fighters of the Pacific Air Forces.

On 31 May, 1967, a KC-135A crew from the 902nd ARefS, home based at Clinton-Sherman AFB, Oklahoma, was involved in a complex and spectacular refueling operation over the Gulf of Tonkin. Their aircraft was KC-135A, 60-3029. Beginning as a routine mission to support a pair of F-104s, it evolved into a save of six fuel-starved carrier-based U. S. Navy aircraft. These included a pair of Douglas KA-3 Skywarriors, two Vought F-8 Crusaders, and another pair of McDonnell F-4 Phantoms. At one point, the KC-135 was refueling a disabled KA-3 which in turn was refueling an F-8. The KC-135 was so depleted of fuel that it had to recover at an alternate base. For their efforts, aircraft commander Major John H. Casteel and his crew of three were awarded the Mackay Trophy. In 1968, SAC's KC-135As supported tactical fighters being deployed in response to the USS Pueblo incident.

This 1967-vintage photograph reveals a pristine KC-135A, 60-0315, refueling a flight of Republic F-105Ds enroute to targets in North Vietnam. This tanker has since been converted to a KC-135R. (USAF SP-82-4425-67[69-022])

Accomplishments of the KC-135As during the war in Southeast Asia are phenomenal. A summary of operations between 1964-1973 follows:

Flying Hours:	911,364
Sorties:	194,687
Refuelings	813,878
Fuel Offload	8,963.7 billion pounds

A total of four KC-135As were lost during operations in SEA, resulting in twenty-three fatalities.

Tanker crews could not buy drinks in the clubs, because of their heroic actions in combat. Instead, the fighter jocks bought all of the rounds in appreciation for those actions which often saved both aircraft and pilot. Tankers routinely went in harms way to assist fuel-starved aircraft. On many occasions, tankers latched onto damaged fighters and dragged them to a friendly base where they could make a dead-stick landing.

KC-135A, 57-1445, was one of several transferred to the 160th ARefG, Ohio ANG, at Rickenbacker AFB. It was the subject of a Revell 1/144th scale tanker kit. The aircraft has since been converted to a KC-135E.
(160th ARefG, Ohio ANG)

Reserve & Air National Guard Tankers

In July 1975, the first of 128 KC-135As were transferred to sixteen Air National Guard and Air Force Reserve units. The first aircraft transferred was KC-135A, 57-1507, from the 301st ARefS of the 301st ARefW. It was reassigned to the 160th ARefG of the Ohio Air National Guard. The transfer took place at Rickenbacker AFB, where both units were co-located.

The 128th KC-135A, 57-1438, was transferred from the 7th BMW at Carswell AFB, Texas, to the 931st ARefG (AFRes) at Grissom AFB, Indiana, on 16 September, 1978. More are transferring all the time.

Another Mackay Trophy

On 5 September, 1983, a routine fighter deployment to Europe was being conducted. While over the Atlantic Ocean, one of the F-4Es developed engine problems. The Phantom diverted to the Gander International Airport, Newfoundland. One engine was shut down and the second began losing power.

A KC-135A, piloted by Major Robert J. Goodman, diverted to assist the floundering fighter. The boom operator provided guidance as the tanker dropped back towards the F-4. Four hookups were made before the fighter was securely latched, thus enabling the tanker to tow it to Loring AFB, Maine. For their actions, Major Goodman's crew was awarded the Mackay Trophy. Both crews were also honored at the Air Force Association's National Convention the following year. Adding insult to injury, the F-4 crew was flown to Europe to rejoin their unit aboard Major Goodman's KC-135A!

Operation El Dorado Canyon

In April 1986, President Ronald Reagan wanted to demonstrate to Colonel Mohomar Kadaffy that he could not continue his international terrorist activities, so Reagan ordered a night air strike on Libya. International agreement on overflights could not be arranged, however England's Prime Minister, Margaret Thatcher, permitted the use of British bases for the operation that was named El Dorado Canyon.

The U. S. Navy, operating from carriers in the Mediterranean, would strike coastal targets, while the U. S. Air Force would venture in with F-111s that would launch from England. So secret were the orders that they were hand-carried to England by senior staff officers. The United States has on-going agreements with Britain for a certain number of no-notice, radio-silent operations per year. Operation El Dorado Canyon would become one of them.

Both KC-10s and KC-135s were prepositioned at British bases. On bases with parallel runways, the tankers would launch from one runway and the fighters would take off from the other. At bases with only one runway, the tankers took off from one base while the fighters departed from a neighboring base using the same tower for the take off call!

On 14 April, the first flights departed between 1800 and 1900 hours. The aircraft flew all night down the Atlantic coast to the entrance to the Mediterranean Sea. Several refuelings were made enroute. Fuel was topped off, and the F-111s headed inland for their assigned targets. One F-111 was lost when heading back out to the coast. The tankers then refueled the F-111s on their return route back out of the Mediterranean and up the Atlantic coast to England.

Operations Desert Shield & Desert Storm

On 2 August, 1990, Iraq's President Saddam Hussein sent his forces charging into Kuwait to reclaim what he called his nineteenth province. He stopped and waited as the UN Coalition force was established. Unbelievably, he allowed this force to mass its resources in the region and train for almost six months!

On 7 August, the 1st Tactical Fighter Wing (TAC) deployed with its armed F-15s from Langley AFB, Virginia. They were supported by KC-135 tankers. By 12 August, thirty-two KC-135s were deployed to bases in Saudi Arabia. SAC eventually deployed a fleet of forty-six KC-10s and 256 KC-135s. Before Christmas, twelve of the thirteen Air National Guard KC-135E units had deployed sixty-two of their aircraft to the theater. These aircraft operated out of eleven different bases in the region.

Operation Desert Storm commenced at 0300, 17 January, 1991. The 2nd BMW launched seven B-52Gs which carried conventionally-armed air-launched cruise missiles. The bombers took off from Barksdale AFB, Louisiana, on 16 January at 0635 CST. This became the longest combat mission ever flown. It lasted more than thirty-five hours, with the targets being communications and power facilities deep inside Iraq. KC-10s and several KC-135s provided refueling support for these B-52s.

A pair of KC-135s sustained major damage during the Gulf War. One aircraft had its belly ripped open by the vertical fin of an undershooting F-111. Both aircraft returned safely to base. A second KC-135 was caught in wake turbulence and two engines were ripped off one side. This aircraft also staggered back to safety. These incidents are a testament to the structural integrity of the basic airplane design. Although the Air Force expected to lose as many as twelve tankers the first night of the war alone, none were lost during the entire conflict.

During the build up under Operation Desert Shield and

A short-tailed KC-135A, 55-3127, is shown here in its original natural metal finish. The wing inspar surfaces were painted with Corogard silver/gray paint. Wing walk areas are well defined in this view, but note that only the red turbine warning stripes were applied. The celestial observation windows are visible in this photograph.
(Boeing-Witchita C1005-1)

later in the combat resupply efforts for Operation Desert Storm, KC-135s provided air refueling support for deploying combat aircraft and MAC's beleaguered force of Lockheed C-141B Starlifters and C-5B Galaxies. During Operations Desert Shield and Desert Storm the forty-six McDonnell Douglas KC-10s and 256 KC-135 tankers accomplished the following:

Desert Shield

Sorties	Hours	Receivers Refueled	Fuel Offload*
4,967	19,089	14,588	68.2

Desert Storm

Sorties	Hours	Receivers Refueled	Fuel Offload*
15,434	59,943	45,955	110.2

* Millions of gallons

Another Record

The U. S. Air Force's term "Global Reach" was re-emphasized in December 1992, when a KC-135R from the 97th Air Mobility Wing made a nonstop, unrefuelled flight between Kadena AB, Okinawa and McGuire AFB, New Jersey. The trip of more than 8,700 nautical miles was made in seventeen hours and thirty-one minutes, and the aircraft landed with 29,000 pounds of fuel which was sufficient for two more hours of flight. The tanker departed Kadena AB at a gross weight of 322,500 pounds which included 195,000 pounds of fuel, 5,000 pounds of mission-essential cargo, and a crew of eleven. On this flight, the crew established a new distance record for aircraft in the 220,460 to 330,690-pound category.

Downsizing the Force

The U. S. armed forces are undergoing a large scale downsizing now that the Cold War has been won. There are members of Congress who believe that we can make Draconian cuts in our military force at a rapid rate, and the Clinton Administration is spearheading this move.

Aerial refueling is a national asset, which along with our strategic bomber force, is being severely reduced. Current plans call for a strategic bomber force of about two hundred aircraft. As of this writing, almost eighty-four -135s of several types are already in storage at Davis-Monthan AFB, Arizona. More are being added on a regular basis. A pair of -135s have been relegated to ground trainer duties, while another ten are on static display at a number of air bases. These include:

TYPE	Serial No.	Display Location
C-135C	61-2671	Tinker AFB, OK
EC-135A	61-0287	Offutt AFB, NE
EC-135A	61-0262	Ellsworth AFB, SD
EC-135C	63-8049	SAC Museum, Omaha, NE
EC-135J	63-8057	Pima Air Museum, Tucson, AZ
EC-135L	61-0269	Grissom AFB, IN
KC-135A	55-3139	Castle AFB, CA
KC-135A	56-3639	Dyess AFB, TX
KC-135A	55-3130	March AFB, CA
NKC-135A	55-3123	USAF Museum, Dayton, OH

Channel Runs

The Gulf War and subsequent humanitarian relief missions took their toll on the aging strategic airlift assets of the U. S. Air Force. The C-5 Galaxies are now undergoing major overhauls earlier than expected, thus making fewer of these large transports available for use. Many of the C-141s are having major structural problems, including cracks in their wings and leaks in their integral fuel tanks. These problems are causing the aircraft to be grounded or accelerated through depot maintenance. As a result, there is a major shortfall in available transports to carry the myriad of spare parts and other high-volume, low-weight priority cargo.

With the reorganization of the Air Force in the 1991-1992 time frame, Military Airlift Command (MAC) was disbanded and replaced by the new Air Mobility Command (AMC) which took the strategic airlift fleet of MAC and the tanker assets of the former Strategic Air Command (SAC). To fill the gap in strategic airlift, the Air Force has fitted a number of KC-135s with components of the 463-L cargo roller system (used in the C-5 and C-141), so that the tankers can perform these transport missions which are known as Channel Runs. KC-135s from AMC, the Air National Guard, and the Air Force Reserve are being used in this capacity. Each aircraft can carry up to fifteen tons of cargo, Should a contingency arise, the KC-135s will revert back to their tanker mission.

Proven Value

Since the end of World War II, few military aircraft programs can compare to that of the -135 for speed, execution, magnitude, cost effectiveness, and operational success. These airplanes were delivered at the lowest cost-per-pound of airframe weight of any production military airplane program at that time. For the total production run of 820 aircraft between 30 April, 1957 and 12 January, 1965, the total cost was approximately $1.66 billion, at an average price of $2 million per airframe.

The KC-135 has proven itself in training and combat. Its basic design was so versatile that a large number of derivatives were possible. Through its myriad of upgrades, the aircraft is scheduled to remain in active service well into the twenty-first century. The -135 family has exceeded expectations and has proven to have been an excellent value to the American taxpayer.

PRODUCTION BLOCKS AND SERIAL NUMBERS
(AS DELIVERED)

Serial Number	Block Number
56-3591/56-3606	KC-135A-05-BN
56-3607/56-3631	KC-135A-06-BN
56-3632/56-3658	KC-135A-07-BN
57-1418/57-1460	KC-135A-08-BN
57-1461/57-1495	KC-135A-09-BN
57-1496/57-2606	KC-135A-10-BN
58-0001/58-0034	KC-135A-11-BN
58-0035/58-0064	KC-135A-12-BN
58-0065/58-0095	KC-135A-13-BN
58-0096/58-0130	KC-135A-14-BN
59-1443/59-1472	KC-135A-15-BN
59-1473/59-1502	KC-135A-16-BN
59-1503/59-1523	KC-135A-17-BN
60-0313/60-0342	KC-15A-18-BN
60-0343/60-0368	KC-135A-19-BN
60-0369/60-0378	C-135A-01-BN
61-0261/61-0277	KC-135A-19-BN
61-0278/61-0302	KC-135A-20-BN
61-0303/61-0325	KC-135A-21-BN

Serial Number	Block Number
61-0326/61-0330	C-135A-02-BN
61-0331/61-0332	C-135B-01-BN
61-2662/61-2666	C-135B-02-BN
61-2667/61-2674	C-135B-03-BN
62-3497/62-3519	KC-135A-22-BN
62-3520/62-3548	KC-135A-23-BN
62-3549/62-3580	KC-135A-24-BN
62-3581/62-3585	KC-135B-01-BN
62-4125/62-4139	C-135B-04-BN
63-7976/63-7980	KC-135A-24-BN
63-7981/63-8015	KC-135A-25-BN
63-8016/63-8045	KC-135A-26-BN
63-8046/63-8057	KC-135B-01-BN
63-8058/63-8061	RC-135A-01-BN
63-8471/63-8475	C-135F-00-BN
63-8871/63-8888	KC-135A-26-BN
63-9792	RC-135B-01-BN
63-12735/63-12740	C-135F-00-BN
64-14828/64-14840	KC-135A-27-BN
64-14841/64-14849	RC-135B-01-BN

KC-135 FAMILY TREE

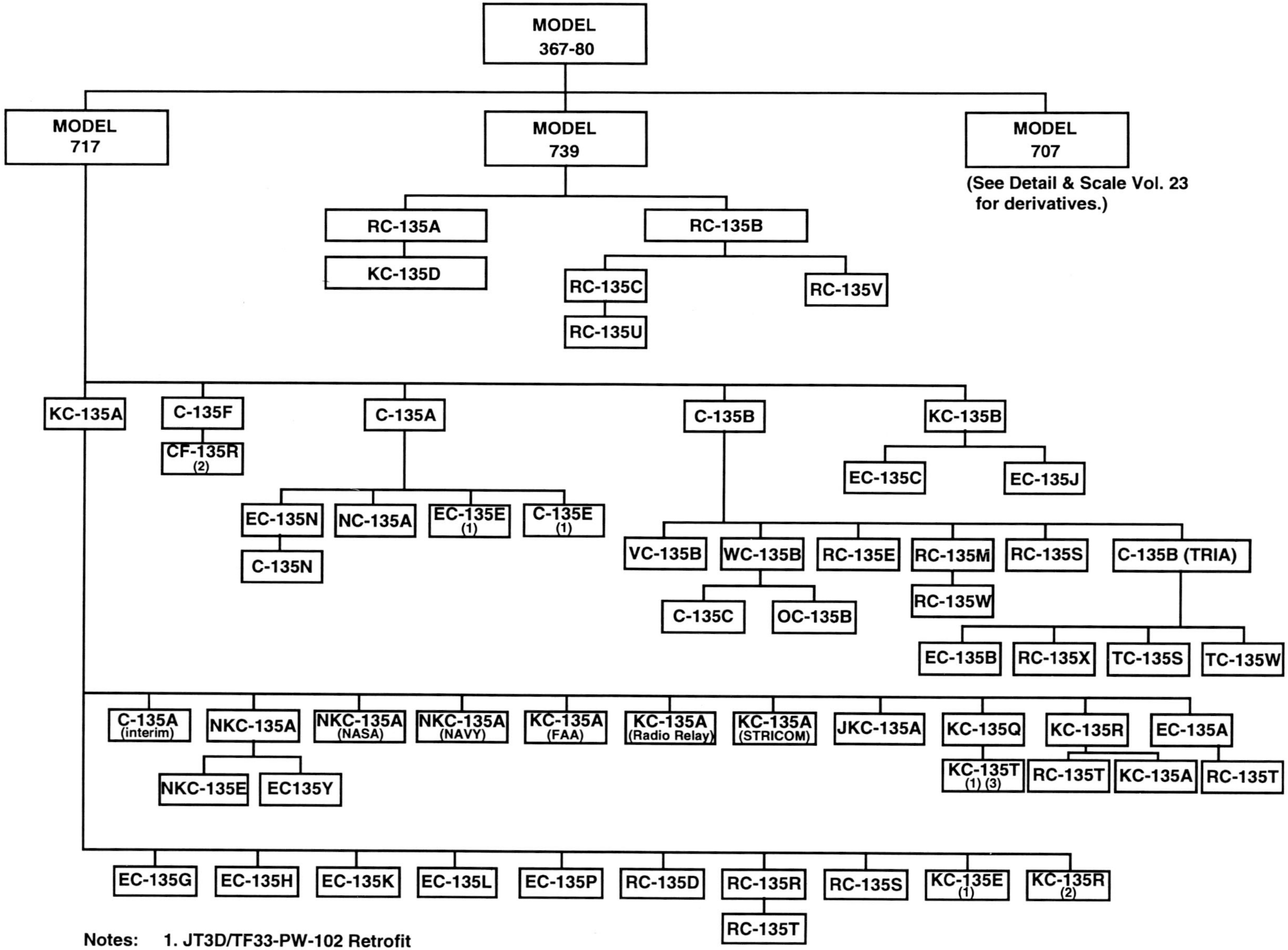

Notes: 1. JT3D/TF33-PW-102 Retrofit

2. CFM-56 Retrofit

3. Subject to USAF reorganization with KC-135Q mission going from ACC to AMC

DIMENSIONS

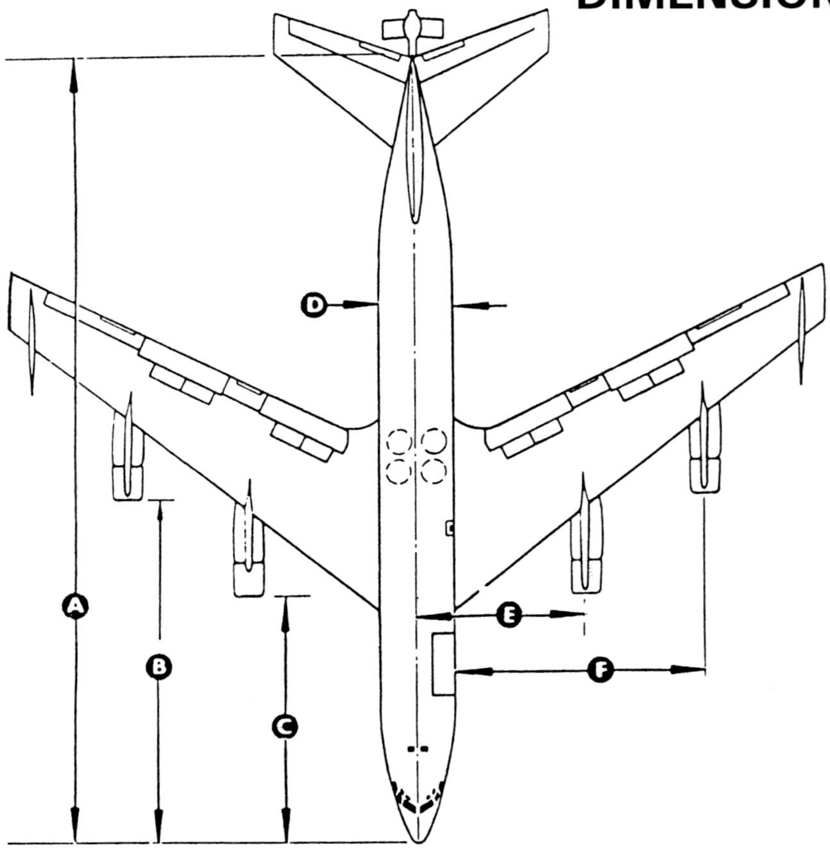

WING AREA..... 2433 SQUARE FEET
SWEEP BACK .... 35 DEGREES (1/4 CHORD LINE)
DIHEDRAL....... 7 DEGREES

DIMENSION	KC-135A	KC-135B	C-135A	C-135B	RC-135A	RC-135B
A	128' 10"	128' 10"	128' 10"	128' 10"	128' 10"	128' 10"
B	56' 3"	55' 7"	56' 3"	55' 7"	56' 3"	55' 7"
C	41' 0"	40' 4"	41' 0"	40' 4"	41' 0"	40' 4"
D	12' 0"	12' 0"	12' 0"	12' 0"	12' 0"	12' 0"
E	27' 2"	27' 0"	27' 2"	27' 2"	27' 2"	27' 2"
F	46' 1"	46' 1"	46' 1"	46' 1"	46' 1"	46' 1"
G	130' 10"	130' 10"	130' 10"	130' 10"	130' 10"	130' 10"
H	39' 8"	43' 0"	39' 8"	43' 0"	39' 8"	43' 0"
I	22' 1"	22' 1"	22' 1"	22' 1"	22' 1"	22' 1"
J	134' 6"	135' 1"	134' 6"	135' 1"	134' 6"	135' 1"
K	[1]	41' 8"	41' 8"	41' 8"	41' 8"	41' 8"
L	13' 10"	13' 10"	13' 10"	13' 10"	13' 10"	13' 10"
M	17' 10"	17' 10"	17' 10"	17' 10"	17' 10"	17' 10"
N	45' 8"	45' 8"	45' 8"	45' 8"	45' 8"	45' 8"

[1] AIRPLANES WITH POWER RUDDER 41'-8"
AIRPLANES WITHOUT POWER RUDDER 38'-5"

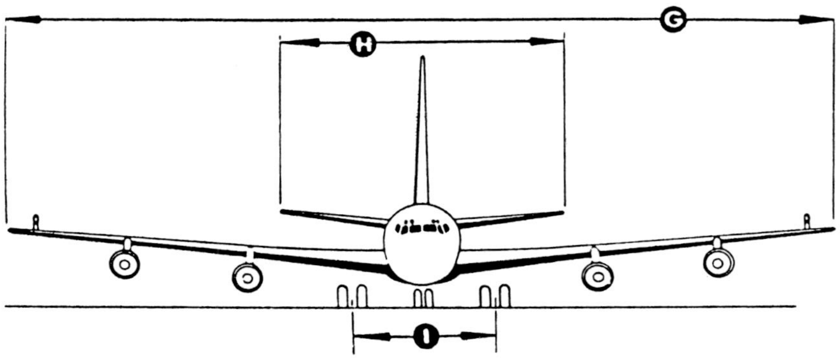

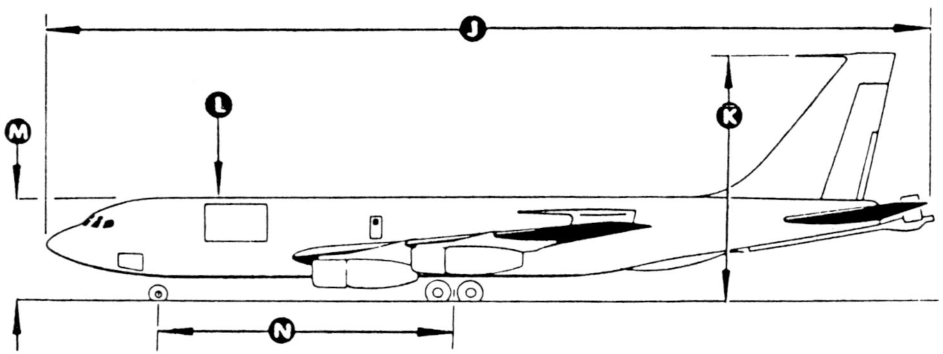

KC-135A FIVE-VIEW SCALE DRAWINGS

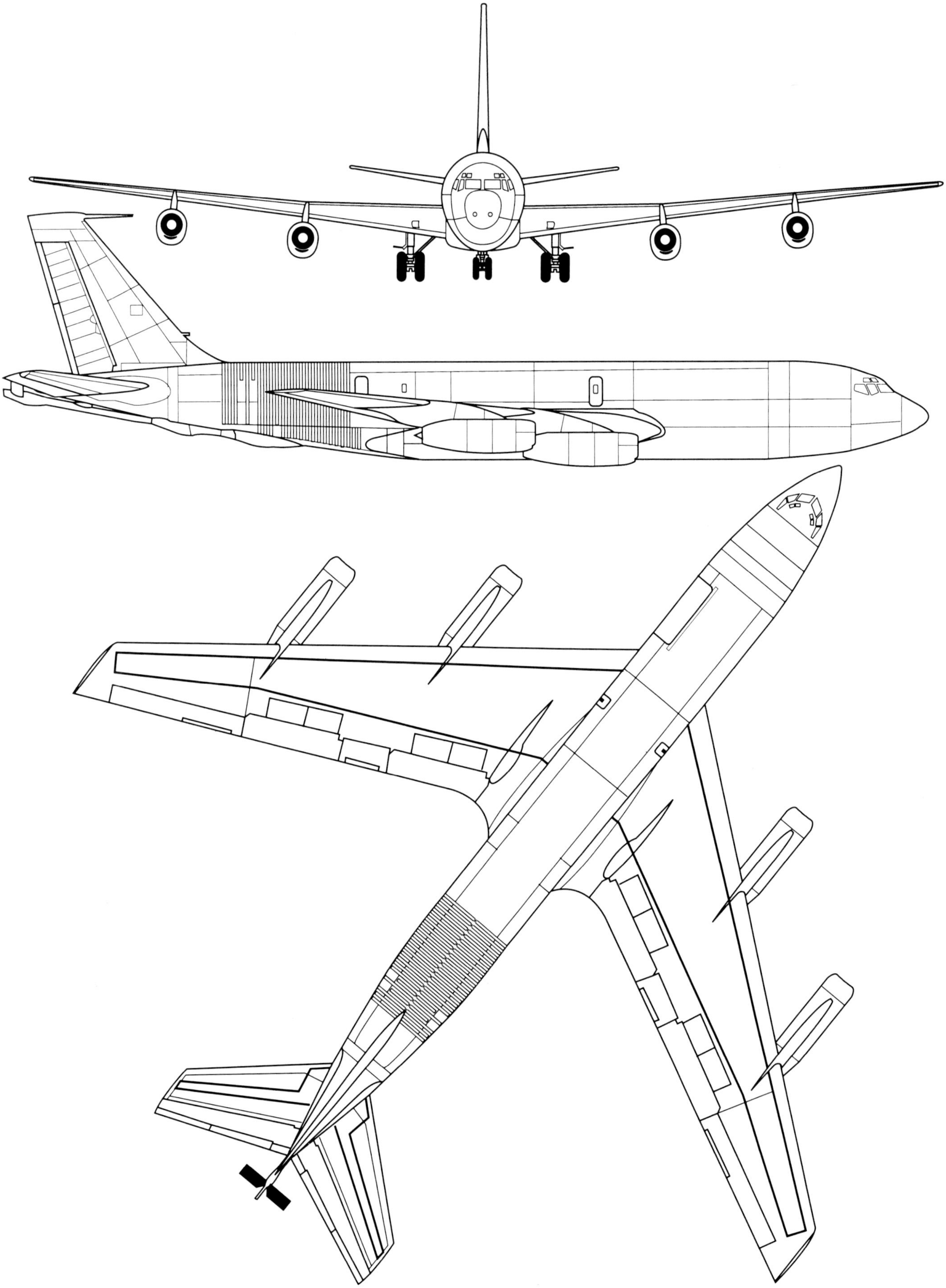

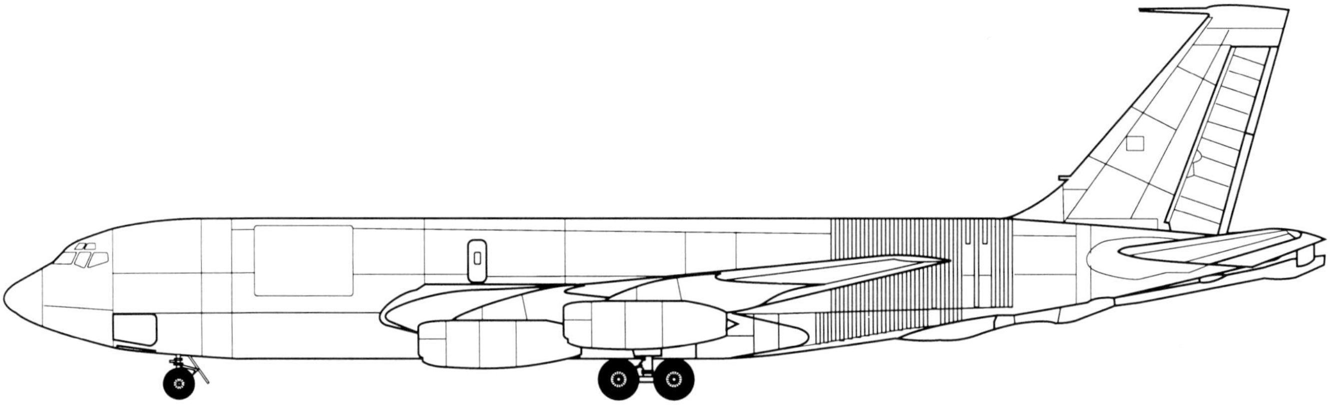

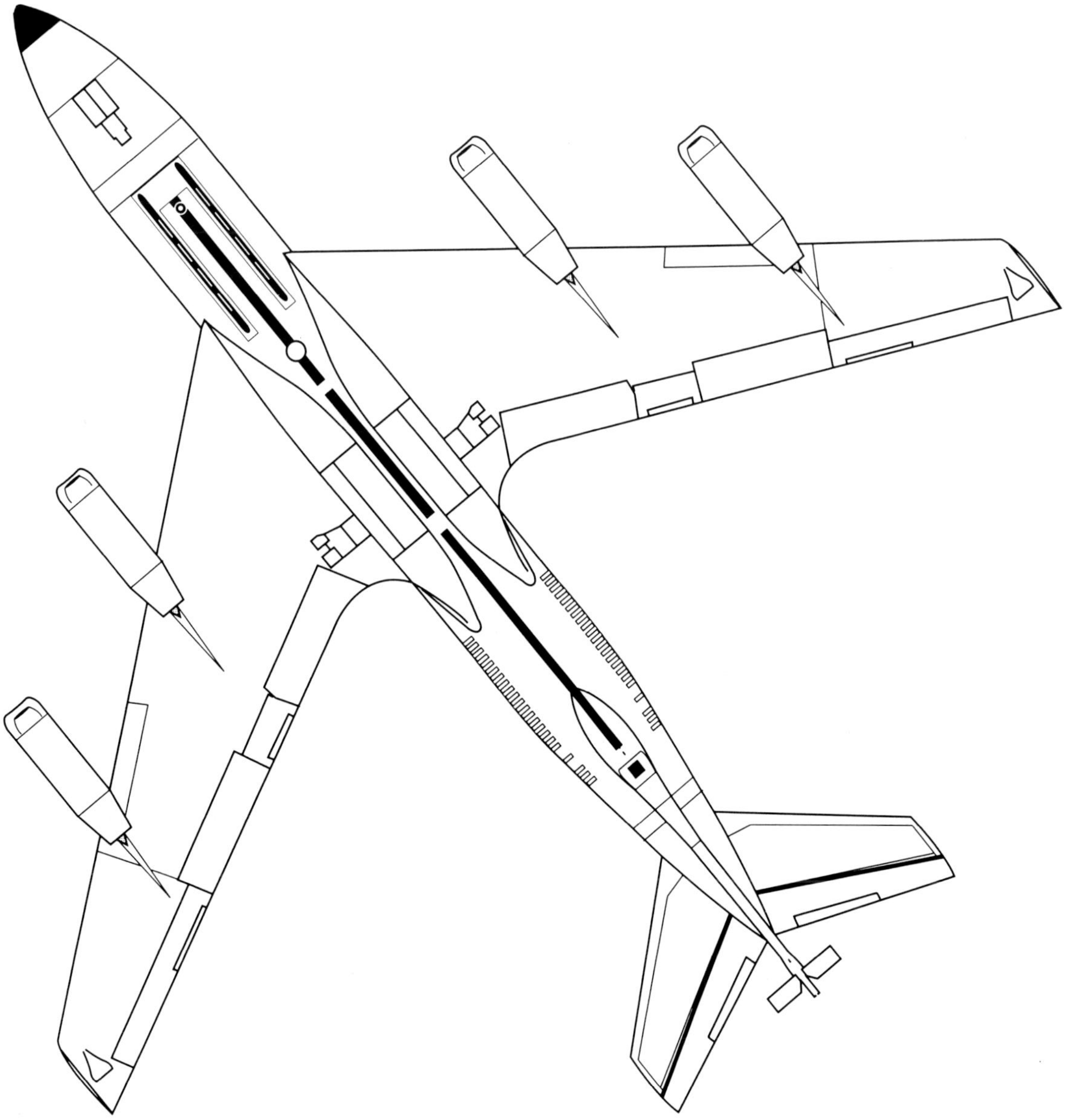

AIRPLANE DESCRIPTION & DETAILS

The KC-135A was designed to meet the air refueling and transport needs of Strategic Air Command (SAC). The lower fuselage and wings carry both air refueling and aircraft fuel. The main cabin area is available for transporting troops and/or cargo.

STRUCTURES

The semi-monocoque fuselage may be pressurized in all areas except the wheel wells, body fuel tank areas in the lower compartment, and the area to the rear of the aft pressure bulkhead at Body Station 1440. The fuselage bulkheads and frames are spaced at twenty-inch intervals. Longitudinal stringers are attached to the bulkheads and frames. Most of the skin panels are butt-joined to maintain aerodynamic smoothness. A hinged crew access door is located on the left side of the forward fuselage near the nose wheel well, and three escape hatches are in the main cabin area. One of these is on the left side of the overwing area and a second is directly opposite it on the right side. The third hatch is on the right side of the fuselage just forward of the boom operator's station. A main cargo door, which is 72 by 144 inches in size, is located on the left side of the fuselage forward of the wing. Two main windshields, four cockpit side windows, four eyebrow windows, the windows in the boom operator's station, and the windows in the exit hatches comprise the standard transparencies on the airplane. Additional windows have been added on specialized variants.

The wing is a semi-monocoque, fully cantilevered, cellular structure tapering in both planform and depth. A sweepback of thirty-five degrees at the quarter chord line is employed. The wing consists of five spanwise divisions, including the wing center section, a pair of inboard panels, and two outboard panels. The inspar wing box forms an integral wing tank except in the center section which houses a bladder cell. With the exception of the tips, the entire trailing edge of the wing is made up of control surfaces. Hinged cove panels on the wing's lower surface allow maintenance access to the flaps and wing rear spar and maintain laminar airflow in flight when the flaps are retracted. The wings are mounted to the fuselage with a dihedral of seven degrees.

The horizontal stabilizer is a two-spar, swept structure which pivots about the aft spar to provide pitch control. Like the wings, the horizontal stabilizer is installed with a dihedral of seven degrees. A series of vortex generators is mounted on the lower surface to assure laminar airflow for the elevator. Three different horizontal stabilizer and elevator configurations have been installed on the -135 family. The -135s delivered with J57 turbojets had a stabilizer and elevator spanning 39′ 8." Aircraft delivered with production TF33 turbofans had elevators spanning 39′ 8," but the standard stabilizer had a 24.27-inch extension between the elevator and stabilizer end cap. The first eighty -135 aircraft that were retrofitted with cannibalized JT3D (TF33-PW-102) engines had the same horizontal stabilizers as the jetliners. These had the original elevators. All subsequent JT3D and all CFM56 retrofits had their existing stabilizers modified with the 24.27-inch tip extensions.

The vertical fin is also a two-spar structure. Early KC-135As were delivered with short vertical stabilizers. The height of these aircraft was thirty-eight feet, five inches. A curved fin cap was mounted directly above the fin and rudder. When the powered rudder was installed on the 583rd airplane (serial number 62-3532) a forty-inch extension was added to the top of the vertical stabilizer thereby increasing the surface area and aspect ratio.

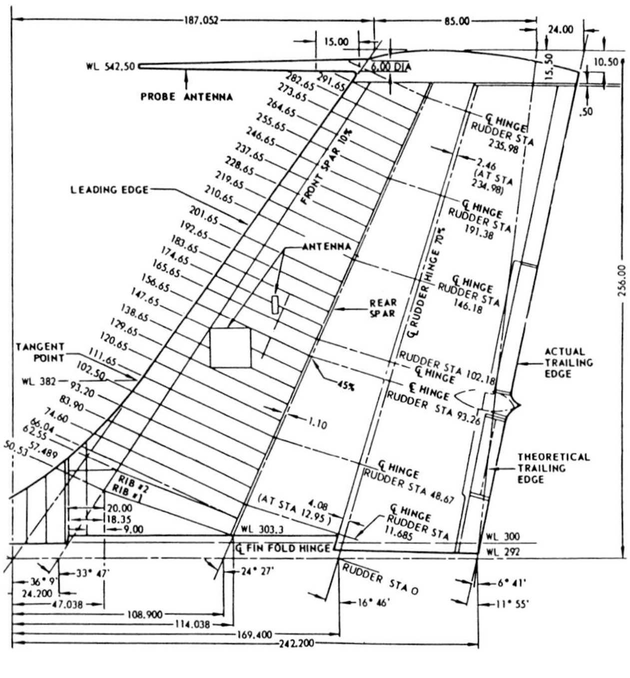

The original vertical fin was 256.00 inches high.

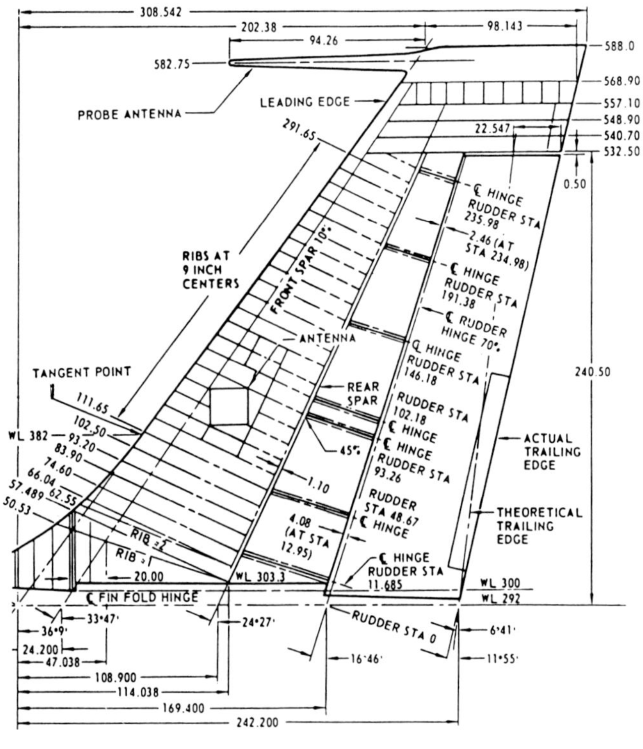

A forty-inch extension was later added to the vertical fin and was retrofitted to existing aircraft.

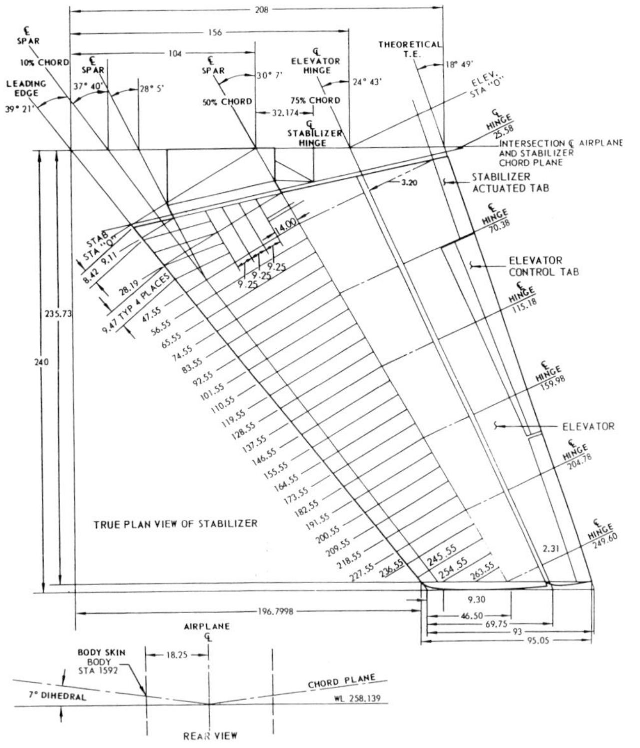

All J57-powered -135s were delivered with this horizontal stabilizer which measured 340 inches in span.

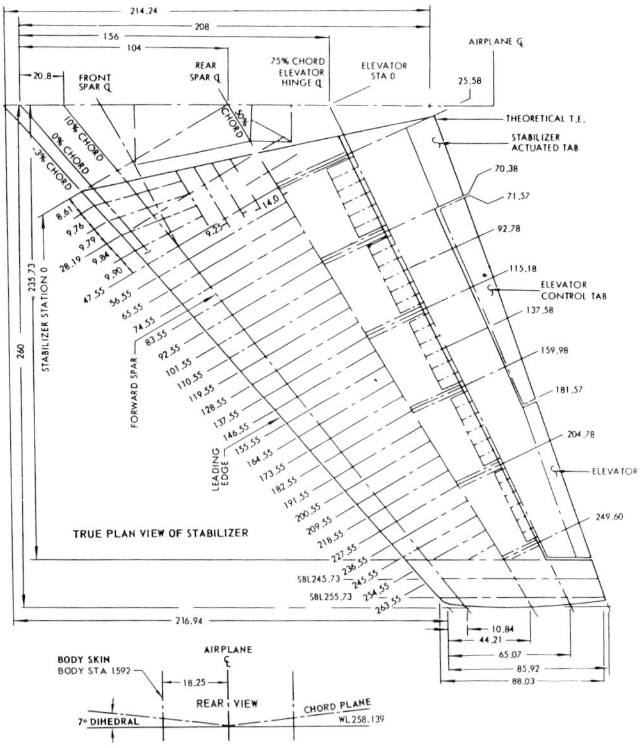

TF33 fan-powered -135s were delivered with horizontal stabilizers that were fitted with a twenty-inch tip extension.

Airplane height was increased to forty-one feet, eight inches. The vertical stabilizer is hinged to fold to the right for hangar storage and maintenance, hence the walkway markings are on the left surface.

ENVIRONMENTAL CONTROLS

The environmental control system is divided into four systems. These include the pneumatic bleed air system, the cabin pressure control system, the cabin air conditioning system, and the oxygen systems.

Pneumatic Bleed Air System

The pneumatic bleed air system consists of four duct runs, one from each engine, providing air to the manifold in the wing leading edge, and in turn to the engine bleed air crossover valve in the fuselage. This air then enters the air conditioning system. Ninth stage compressor air is tapped on the J57 engines, whereas two compressor bleed sources (low and high stage) are employed on the turbofan engines.

Cabin Air Conditioning System

The air conditioning system is located on either side of the keel beam. The system consists of a pair of air cycle machines which are electrically and pneumatically controlled. Outside air, which is drawn in through the ram air scoops located in the fuselage near the wing leading edge, is warmed with engine bleed air in the heat exchangers. This conditioned air is then directed through a series of manifolds and riser ducts into the cockpit and main cabin.

Cabin Pressure Control System

The pneumatically operated cabin pressure control system pressurizes the cabin. The system is capable of maintaining any cabin pressure differential between 0.2 and 9.42 pounds per square inch.

Oxygen Systems

Four different oxygen systems are employed on the -135 series aircraft. These include an eight-liter liquid oxygen system, a 150 liter liquid oxygen system, an eight cylinder high pressure gaseous oxygen system, and a six cylinder high pressure gaseous oxygen system.

These five yellow cylinders contain aviators' breathing oxygen for the crew. They are installed in the aft cabin above the refueling boom pod. To the rear are a pair of white urinals which are located along the aft bulkhead.

PMEE Cooling System

A supplementary Primary Mission Electronic Equipment (PMEE) Cooling System has been added to the C-135E and EC-135E/N aircraft.

Pressure Equalization System

A Pressure Equalization System has been added to the WC-135B aircraft.

POWERPLANT

Each of the four engines are suspended and cantilevered ahead of the wing by a strut assembly. The pylon assemblies are mounted at a ninety-degree angle to the wing chord plane. A series of three cone bolts attaches each to mount assembly within the strut. Six different types of powerplants have been installed on the -135 family of aircraft. (See the accompanying table for the series -135/engine match-up.) It will be noted that some series have more than one version of the J57 engine series because of the time frame when the aircraft came off the production line.

ENGINE	**AIRPLANE SERIES**
J57-P/F-43WB	C-135A, GNC-135A, GNKC-135A, KC-135A, NC-135A, NKC-135A
J57-P/F-59WP	C-135A, EC-135A, EC-135G, EC-135L, KC-135A, KC-135Q, NKC-135A
TF33-PW-5	C-135B, C-135C, RC-135S, RC-135W, RC-135X, TC-135S, TC-135W, WC-135B
TF33-PW-9	EC-135C, EC-135J, RC-135U, RC-135V
TF33-PW-102	C-135E, EC-135E, EC-135H, EC-135K, EC-135N, EC-135P, EC-135Y, KC-135D, KC-135E, NKC-135E
F-108-CF-100	C-135FR, KC-135R

Engine Descriptions:

Pratt & Whitney J57-P/F-43WB and J57-P/F-59W Turbojet

These engines are rated in the 11,200-pound static thrust class and are equipped with water injection which increases the thrust to 13,750 pounds for takeoff. (Hence the heavy black smoke during takeoff.) Engines Number 1, 2 and 3 are equipped with an AC generator; while the Number 4 engine is equipped with a cartridge starter. The engine instruments consist of an exhaust gas temperature (EGT) gauge, engine pressure ratio (EPR) gauge, tachometer, fuel flow meter, and oil pressure gauge. Compressor high stage bleed air provides anti-icing for the engine inlet cowl, nose dome and Pt_2 probe. The major difference between the -43 and -59 engines is the fuel control. The -43 has JFC12-3 fuel controls and Chandler-Evans fuel pumps, while the -59 engines have AJ-A4 fuel controls and Pesco fuel pumps installed.

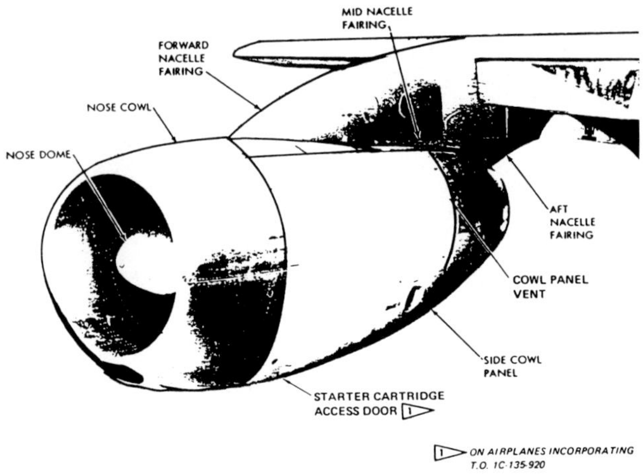

The typical J57 engine cowling is illustrated in this drawing.

Pratt & Whitney TF33-PW-5 Turbofan

TF33-PW-5 engines deliver 18,000-pounds of static thrust, and are two-stage turbofans which provide about seventy-five percent of their thrust from the turbine and twenty-five percent from the fan. These engines were initially delivered on the KC-135B, RC-135B and C-135B aircraft. A cartridge/pneumatic starter system is provided for the Number 3 engine, while the remaining engines have only a pneumatic starter. Compressor high stage bleed air provides anti-icing for the engine inlet cowl, nose dome and Pt_2 probe. Thrust reversers divert the flow of fan discharge air and deflect the engine exhaust gasses. The engine instruments consist of an EGT gauge, EPR gauge. N_1 and N_2 tachometers, an oil pressure gauge, and an oil temperature gauge. None of these airplanes have strut humps, because their forty-KVA generators are located on the engine accessory drive.

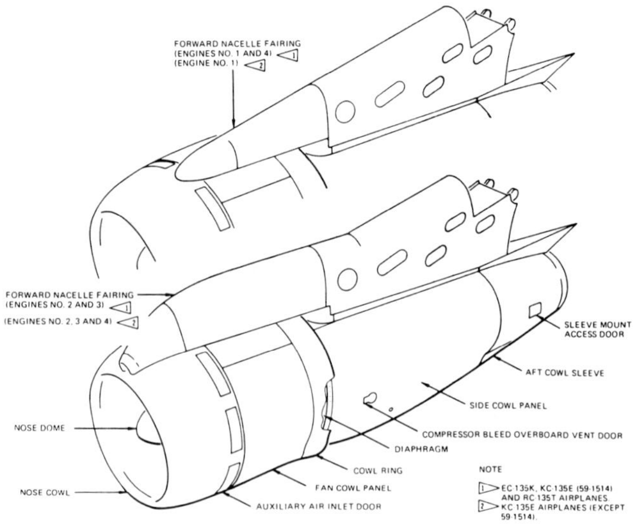

The TF33 turbofan engine cowls are shown here with both the humped and unhumped struts. The generator drive is housed in the humped fairing on RC-135s, but it does not appear on the WC-135Bs or on any of the MATS/MAC/AMC -135 transports.

Pratt & Whitney TF33-PW-9 Turbofan

These engines are in the 18,000-pound static thrust class and are two-stage turbofans which are similar to the TF33-PW-5. However, they are not equipped with thrust reversers. Because they were installed on KC-135Bs and RC-135Bs, there was a requirement for 120-KVA electrical generators to be located within the nacelle struts. Therefore, the struts are humped.

Pratt & Whitney TF33-PW-102 Turbofan

TF33-PW-102 engines also produce 18,000 pounds of static thrust and are commercial JT3Ds that were taken from decommissioned Model 707 jetliners. They have been retrofitted on Air National Guard and Air Force Reserve KC-135As, thus producing the KC-135E variant. Additionally, they have also been used on a number of special purpose aircraft. These engines differ from the TF33-PW-5 engines in that they have an additional compressor stage. Because they came from commercial jetliners, there are a variety of pylon fairings accompanying them. Some aircraft have two or three engine struts with the humped leading edge fairings. This depends upon whether they came from 707-100B/720B or 707-300B/C airplanes. On commercial jetliners these humped fairings provided inlets for the turbocompressors associated with the cabin pressurization system.

General Electric/SNECMA F-108-CF-100 (CFM56-2) High Bypass Thrbofan

These are the most powerful engines used on any versions of the -135 family of aircraft, and they produce up to 20,000 pounds of static thrust. They have been retrofitted on a number of KC-135As, and this resulted in those aircraft being redesignated as KC-135Rs. In addition to producing greater thrust, these engines have a lower specific fuel consumption (SFC), thereby greatly enhancing their efficiency. Another benefit is a markedly decreased noise signature.

All C-135B and RC-135B aircraft were delivered with TF-33 turbofan engines equipped with thrust reversers. All follow-on modifications of these aircraft are so equipped.

The mystery of humped versus unhumped engine struts (those with and without turbocompressors) is revealed in the following table. Note that KC-135E retrofits vary extensively because of the variety of strut configurations coming from donor aircraft.

	Engine Position			
Series	1	2	3	4
C-135B	N	N	N	N
C-135C	N	N	N	N
RC-135E	N	N	N	N
RC-135M	N	N	N	N
RC-135S	N	N	N	N
RC-135W	N	N	N	N
WC-135B	N	N	N	N
OC-135B	N	N	N	N
EC-135C	H	H	N	H
EC-135J	H	H	N	H
RC-135B	H	H	N	H
RC-135C	H	H	N	H
RC-135U	H	H	N	N
RC-135V	H	H	N	H (*)
KC-135E	N	H	H	N (**)
	N	H	H	H (***)
	H	H	H	H (***)

N = No Fairing
H = Humped Fairing
* = tail no. specific
** = usual
*** = sometimes

KC-135Rs are powered by these large F-108-CF-100 (CFM56-2) high-bypass engines.

(Boeing-Wichita C882-585)

The F-108 engines are installed tightly against the wing and are cantilevered forward to afford adequate ground clearance. The warning stripes are to alert the ground crews about the hazzards of being ingested into the compressor section or burned by the exhaust. Note the red rubberized canvas boot which is held snugly in place with elastic straps.

FUEL SYSTEM

The fuel system consists of a number of major components including an engine fuel feed system, fueling and defueling system, gravity transfer and fuel dump system, wing tank vent system, body tank vent system, bladder cell cavities and drains, reserve fuel tanks, main wing tanks, center wing tank, body fuel tanks, fuel quantity indicating system, and a water injection system (on the C/KC-135As only).

The main wing tanks, reserve fuel tanks, and wing center tanks are generally used to carry fuel that will be used by the tanker. The forward and aft body fuel tanks are used to service the receiver aircraft. All onboard fuel may be used by the tanker; whereas all but 1,000 gallons may be used for refueling. The following fuel tank capacities are available:

TANK	AIRPLANE SERIES			
	Basic		RC-135S/W	
	Gallons	Lbs	Gallons	Lbs
Main Wing No. 1	2,113	13,734	2,113	13,734
Main Wing No. 2	2.113	13,734	2.113	13,734
Main Wing No. 3	2,279	14,814	2,279	14,814
Main Wing No. 4	2,279	14,814	2,279	14,814
Reserve No. 1	434	2,821	434	2,821
Reserve No. 4	434	2,821	434	2,821
Center Wing	7,271	47,262	7,271	47,262
Fwd Body	5,800	37,700	2,996	19,474
Aft Body	6,378	41,717	6,378	41,717
Upper Deck*	2,174	14,131	-	-

* KC-135A/KC-135D

The single-point refueling panel and receptacle are located along the keel beam in the forward section of the right wheel well.

AIR REFUELING SYSTEM

The air refueling system consists of the boom operator's pod, refueling boom, hydraulic system, receiver director system, and a fuel supply and dump system.

The boom operator's pod is a feature found on all -135s. On non-tanker aircraft, the pod (though slightly redesigned in external shape) houses electronic gear. The pod features three crew pallets which are often called "ironing boards." The center pallet is for the boom operator, and the other two are for instructors or observers. A pair of ruddevator pedals are located at the forward end of the pod for the boom operator, and a ruddevator control stick and a boom telescoping lever are installed at the aft end. An instrument panel is located at the aftmost end of the pod. A retractable sighting door fairs into the compartment when not in use, and it retracts upward providing the boom operator with an excellent view of the receiver. A periscope can be used to track incoming receivers.

The Boeing-designed refueling boom may be moved vertically and through seventeen degrees of azimuth. Full travel of the boom telescoping extension is about eighteen feet.

From its inception, the KC-135A was designed to operate in the boom/receptacle mode. By 1960, it became apparent that the KC-135 could also be used to support fighter aircraft with probe and drogue refueling systems. A hose and drogue adapter was designed to be attached to the end of the KC-135A boom. A collapsible drogue (basket) was attached to a forty-two-inch hose which in turn was connected to a piece of rigid tubing forty-eight inches long. This entire assembly was attached to the boom adapter. Use of the hose and drogue adapter induces a slight drag penalty which results in a slight increase in fuel consumption during cruise. The penalty increases during orbit and refueling with the boom extended.

Some -135s are equipped with a reverse refueling system which permits the tanker to receive fuel from another aircraft. While not commonly used, this system has permitted -135s to cope with certain in-flight emergencies.

Details of the boom operator's control console are shown here.

This is the boom operator's view with the boom deployed.

Some -135 series aircraft are equipped with an air refueling receiver (ARR) system. This system consists of a pair of slipway doors, the ARR receptacle, ARR fuel manifold, pressure disconnect switch, surge relief valve, purge, and scavenge systems.

The refueling boom is shown in the stowed position on this KC-135A from the Washington ANG.

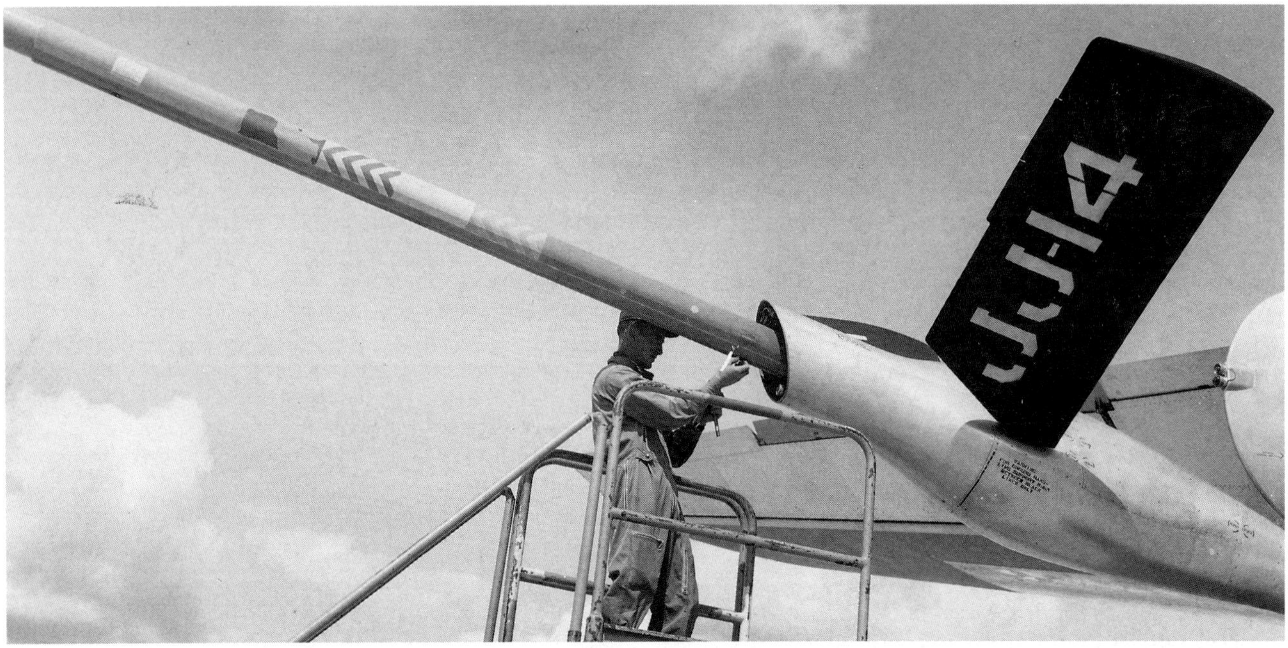

A boom operator inspects the refueling boom. Note the stripes on the telescoping portion of the boom. The code JJ-14 was employed for radio-silent operations. (USAF)

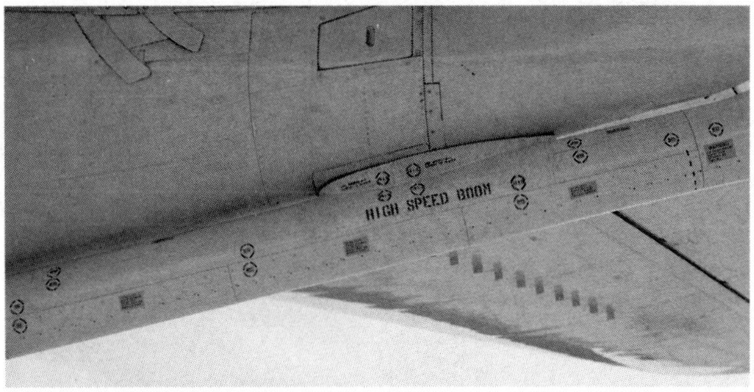

These tufts were added to the fixed boom section to improve aerodynamics and preclude a kick when the boom was in a null position. Note the vortex generators on the lower surface of the horizontal stabilizer.

This is the receiver's view of the KC-135. Note the alignment stripe and the director lights. Lights are also visible within the fixed portion of the boom. *(USAF)*

This view looks forward at the director lights and alignment stripe.

The director lights and alignment stripe are shown here from the front. Note the air conditioning system's ram air inlet under the wing root.

The drogue adapter was attached to this Arizona ANG KC-135A prior to dragging a gaggle of USMC F/A-18 Hornets from Hawaii to California.

The short-legged F/A-18s required five refuelings to get from Hawaii to California even with external fuel tanks.

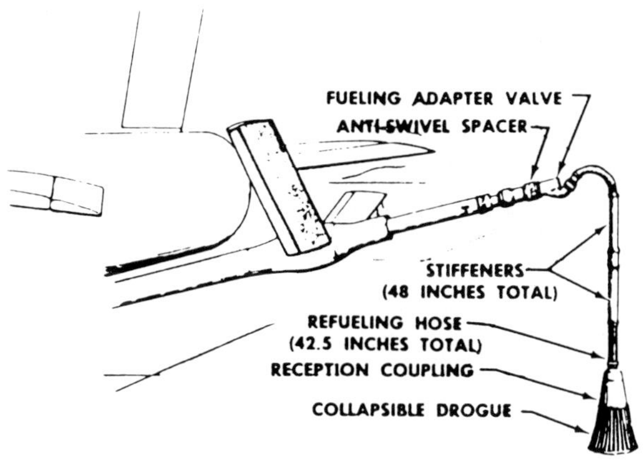

Major components of the drogue assembly are shown here.

A refueling drogue is stowed in the aft cabin of a KC-135.

This view of the nose gear reveals the steering cylinders, upper torsion link, and shock strut door.

LANDING GEAR

The KC-135 employs a retractable tricycle landing gear system. The large main gear assemblies consist of a four-wheel truck which incorporates four axles and two pairs of brake equalizer rods. An anti-skid system is designed into the main gear brake system. The main gear truck is attached to a large clevis on the end of the gear strut inner cylinder. The upper shock strut is attached to the gear trunnion, drag strut, and side strut lower segment. Torsion links connect the inner cylinder to the shock strut.

The nose gear consists of dual axles and wheels which are attached to the inner shock strut cylinder, an outer shock strut cylinder, torsion links, a steering mechanism, trunnion and an extension/retraction system. A shock strut door and taxi light are also attached to the nose gear.

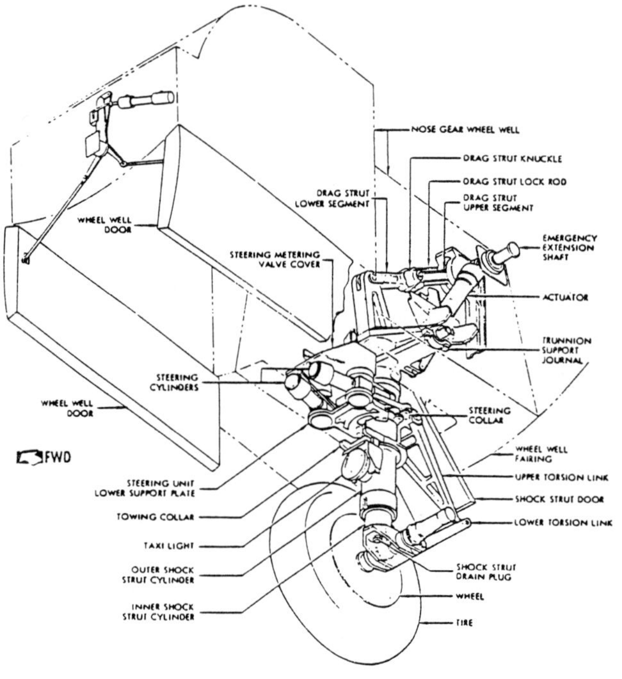

Nose gear system components are shown in this drawing.

The nose gear steeple nestles in the aft portion of the doghouse in the wheel well. The object in the foreground is a disconnected gear door linkage.

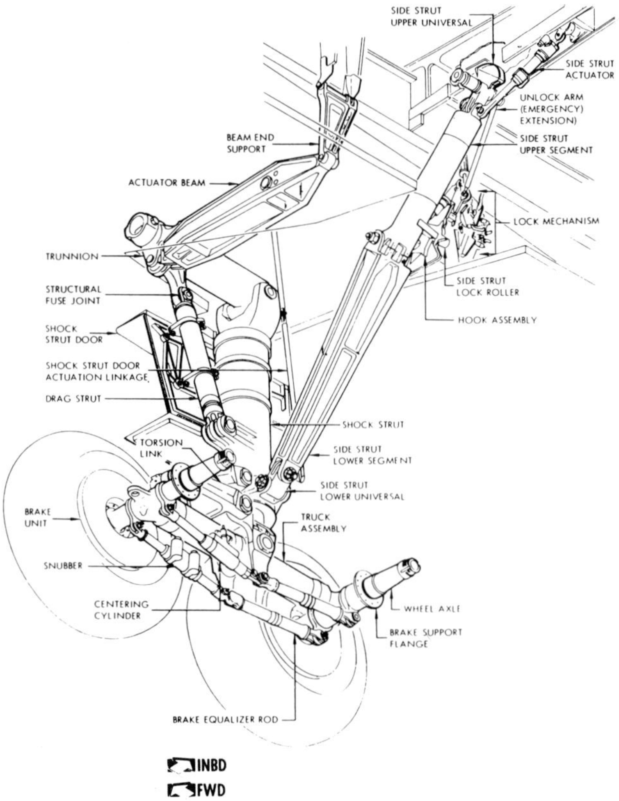

Components of the main landing gear assembly are illustrated in this drawing.

The left main gear assembly is shown here from the rear.

This photo reveals details of the outboard side of the truck assembly on the main gear.

The truck assembly is shown here from the inboard side.

Details of the forward bulkhead of the left main gear wheel well can be seen here. The flap drive gear boxes are yellow, and the large cylinder is a hydraulic accumulator. Note the wheel well door gear lock and the **REMOVE BEFORE FLIGHT** flag. The large pipe elbowing out of the keel beam is for the water injection system.

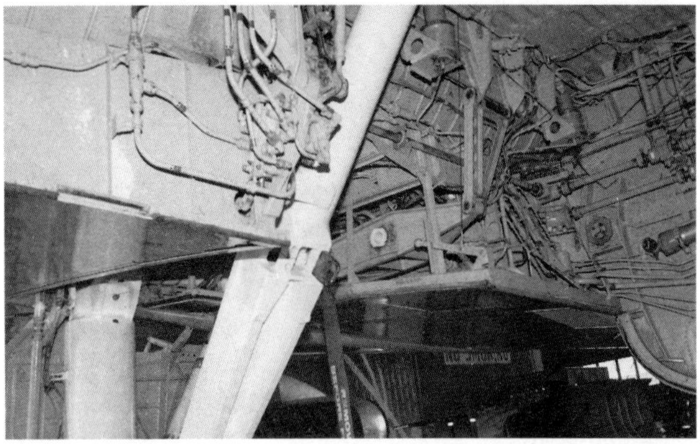

This view looks forward and outward in the left main gear well. All structure and plumbing is painted Aircraft Gray, while the landing gear struts are white. The gear actuating beam and beam link can be seen forward of the gear.

This view looks outboard at the left main gear and reveals the trunnion, drag strut, main strut, side strut, and gear lock at the top left.

KC-135A, 58-0113, was assigned to the 96th BMW when this photograph was taken. Notice the extended trailing edge flaps and the Krueger flaps on the leading edge. (USAF)

FLIGHT CONTROLS

Primary flight controls consist of ailerons, spoilers, elevators, and rudder. These provide control through manual systems which are not hydraulically boosted. Aerodynamic balance tabs assist the flight crew in operating the ailerons, elevators, and rudder. A powered rudder was installed on later production airplanes and retrofitted on all earlier aircraft. Those with boosted rudders may be differentiated by their taller fins. Secondary flight controls consist of Fowler main flaps and split fillet flaps, speed brakes, and an adjustable horizontal stabilizer.

The inboard ailerons are employed throughout all flight speed regimes, whereas the outboard ailerons are used for low speed flight only. A lockout mechanism in the aileron control system disengages the outboard ailerons when the flaps are retracted. The outboard ailerons are moved by direct cable control, not a control tab. A force-relief tab reduces control wheel forces.

Normal flap operation is through the hydraulic system. A manual or electric backup flap operating system is provided. The manual system is on all KC-135As, variants derived from the KC-135A series, as well as RC-135D and RC-135B derivatives. The electric backup system is found on other -135 series.

Low speed directional stability was improved with the introduction of the powered rudder. A forty-inch fin tip extension was added, thereby increasing the vertical stabilizer in area and aspect ratio. The extended fin improved neutral damping action to reduce airplane yaw. The manually-operated rudder on the KC-135As had a maximum deflection of seventeen degrees. The powered rudder has a maximum deflection of twenty-six and one-half degrees when hydraulically operated. Manual reversion offers only seventeen degrees of rudder travel.

Krueger flaps were incorporated along the leading edge of the wing just outboard of the inboard engine pylons. These devices provided additional lift at low airspeeds.

In this view, the rudder is deflected to the left, the horizontal stabilizer is rotated to the upper limit of its green band (takeoff) range, the flaps are extended, and the outboard aileron is in the up position with its control tab in the matching down position. (USAF)

HYDRAULIC SYSTEM

Two independent pressure supply systems provide 3,000 pounds per square inch hydraulic pressure for the -135 aircraft. Engine-driven pumps on the Number 1 and Number 2 engines pressurize the left system, while similar pumps on the Number 3 and Number 4 engines supply the right system.

ELECTRICAL SYSTEM

A 115/200-volt, three-phase, 400-cycle alternating current system and a twenty-eight-volt direct current system make up the basic electrical systems. Three forty-kilowatt volt ampere (KVA) systems powered by constant speed drives (CSDs) on the Numbers 1, 2 and 3 engines provide the AC power on most series of -135s. Some EC/RC-135 series airplanes have three 120-KVA systems. DC power is provided by two transformer rectifiers and a single twenty-eight volt battery.

On fan-powered airplanes, the generator drive oil cooler is prominently located above the fan cowl at the forward end of the engine strut. The small opening at the front of the oil cooler fairing is the air inlet, while the exhaust outlet is located on the left side of the fairing above the fan cowl exit.

Either a Garrett AiResearch JTCP-85 or Solar auxiliary power unit (APU) is installed on production -135s to provide heat and power. The APU is not operable in flight. Fuel for the APU is provided by the aft body tank. A thirty KVA APU system is installed. Airplanes retrofitted with GE/SNECMA CFM56 high bypass turbofans have a new Sunstrand APU installed. The APUs are used for extended ground operation when ground power carts are not available.

AVIONICS

A variety of avionics are common on the -135 fleet. These systems include:

- Interphone system
- Dual VHF-101 command radio systems
- AN/ARC-164 UHF command radio
- Liaison radio system
- Dual ADF system
- Loran navigation system
- Dual 51V-4A glide slope radio systems
- AN/ARN-118(V) navigation system
- Marker beacon radio system
- Radar and rendezvous beacon system
- Search radar system
- N-1 compass system
- J-4 compass system
- Inertial navigation system
- Doppler navigation system (or second INS)
- Dual low range radio altimeter systems
- AN/URT-26(V)-5 crash position indicator system
- AN/APX-64(V) transponder set
- Dual 51R-6 VHF navigation system
- AN/ASQ-141(V) integrated dual flight director system

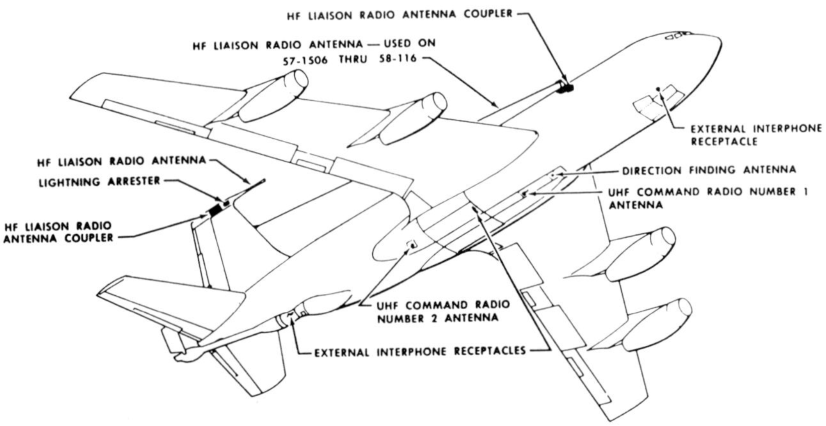

These are the loacations for the external navigation equipment.

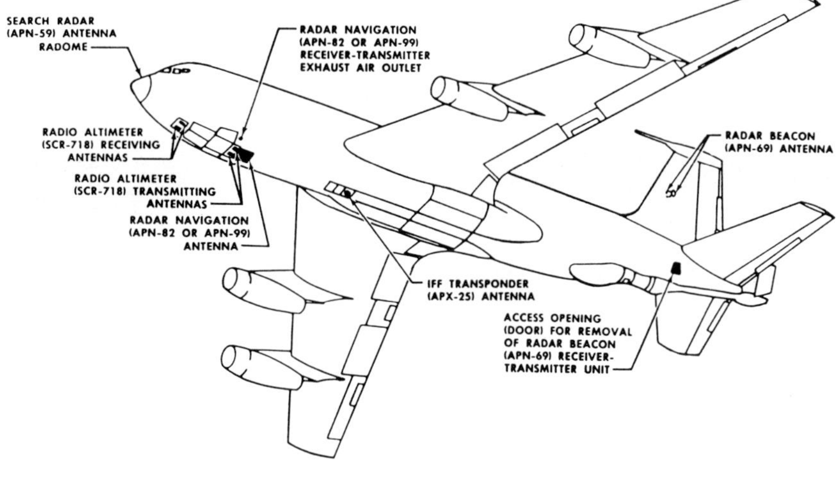

The locations of the external communications equipment are shown here.

EXTERIOR LIGHTING

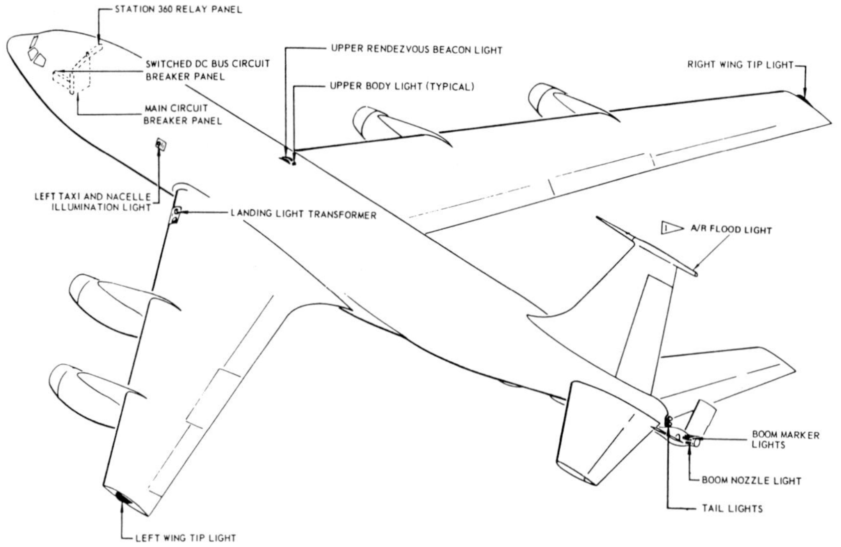

This drawing indicates the various external lights on the upper surfaces of the aircraft. It also points out the locations of the electrical panels in the cockpit.

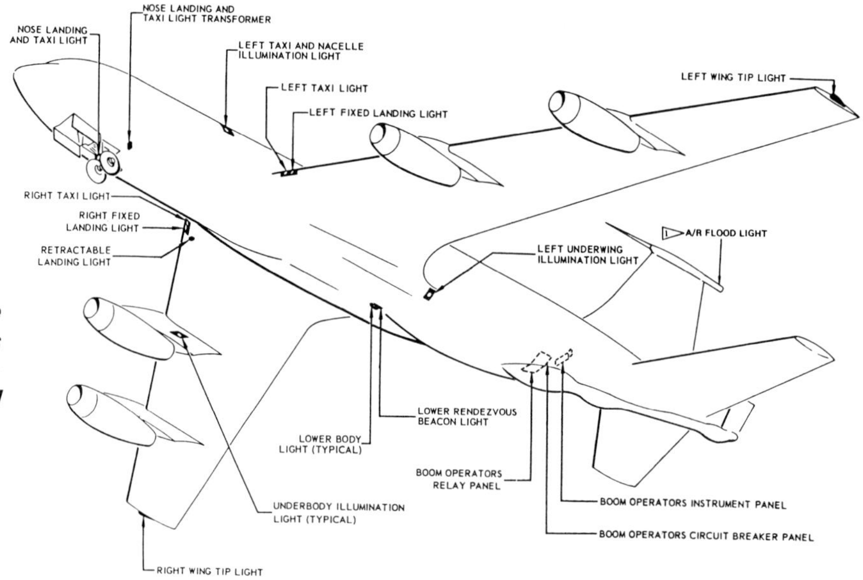

The lights on the undersides of the aircraft are shown in this drawing as is the control panels in the boom operator's pod which have electrical circuits.

Refueling lights are located in the inside aft portion of the sailboat on the inboard strut. Although this is a KC-135R, the lights are similarly placed on all KC-135As.

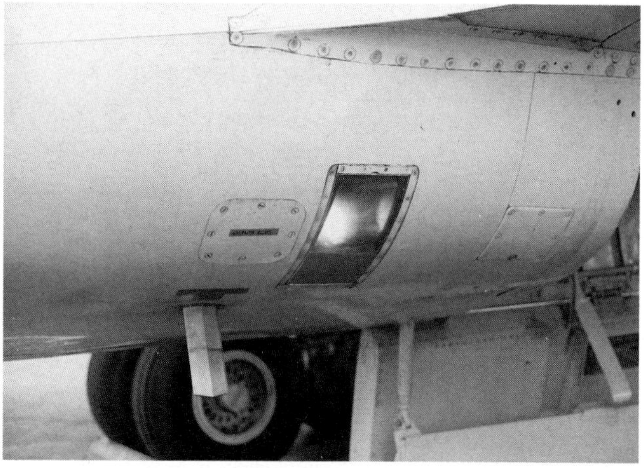

Refueling lights are also located aft of the wheel well for the main gear in the wing-to-body fairing. The tongue hanging out of the aft portion of the wheel well serves as a guide for the wheels as they pass the doors during a manual gear release, thereby preventing the gear from hanging up on the doors.

MISCELLANEOUS DETAILS

The side cargo door can be seen here in its fully extended position. A safety barrier net is installed.

The lower attachment for the crew entry ladder can be seen here.

Here is the side cargo door in a partially open position. Note the two open blow-out panels in the lower portion of the door. When installed, the crew access ladder is attached to the aft bulkhead of the entry hatchway. *(USAF)*

This Air Force Reserve technician is working on the hinge for the forward electronics hatch.

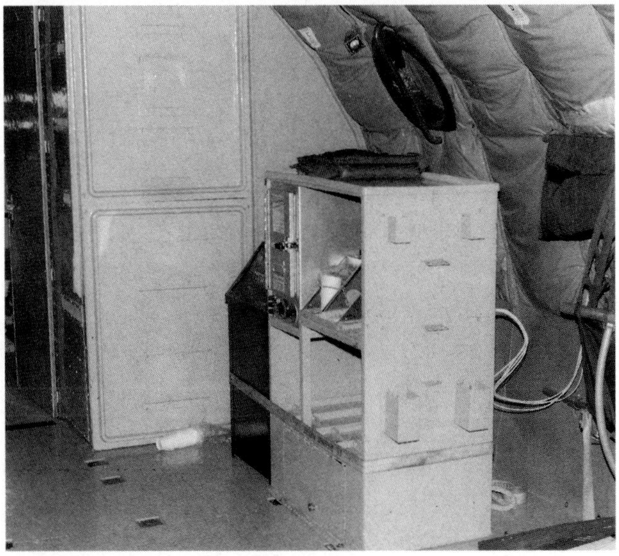

This galley, located opposite the main cargo door, has insert provisions for coffee and water jugs.

Interior details and some of the web troop seats are visible in this photograph. A part of one of the overwing emergency exit hatches is at the extreme right.

This is the boom operator's seat on the flight deck.

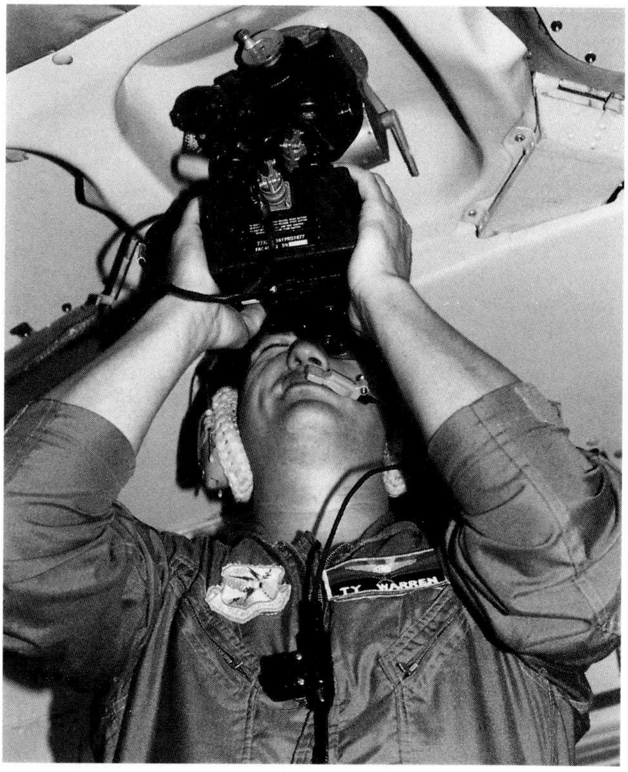

TSGT Ty Warren, from the 161st ARefG, Arizona ANG, assists the navigator with the sun shots for celestial navigation on a flight from Hickam AFB, Hawaii, to Wake Island.

KC-135A

This pristine view of short-tailed KC-135A, 59-1512, reveals the belly bands and skin finish details on a natural metal airplane. This aircraft was delivered to Beale AFB in 1960, and it later became a KC-135Q. (USAF)

Short-tailed KC-135A, 62-3503, was undergoing a control check when this photograph was taken. (USAF)

All of the KC-135As were delivered to the Strategic Air Command with the exception of the early machines, which were part of the test force assigned to the Air Force Systems Command, and three interim transports for MATS. The KC-135As had an air refueling system, which included internal fuel tanks, a boom operator's station, the Boeing-designed flying boom, and a refueling director light system located under the forward fuselage.

The short vertical tail was replaced with one which

Compare the forty-inch fin tip extension on KC-135A, 63-8041, to the fin on the aircraft shown at left. This Stratotanker was later converted into a KC-135R.

Details of the natural metal skin on the lower surface of KC-135A, 59-1512, can be seen here. ***(Boeing P25841)***

was forty inches taller beginning with Line Number 583 (Serial Number 60-3532). A powered rudder, previously developed for the commercial Model 707, was also incorporated. This feature was retrofitted to all earlier -135s.

The standard crew on a KC-135A consists of two pilots, a navigator and a boom operator. Seats for all four are located in the cockpit. Another seat is available for a cockpit observer. This folding seat is stowed behind the copilot and slides out to a position aft and between the pilots' seats. For deployments to bases where KC-135s were not usually assigned, the crew is augmented with two or three crew chiefs and sometimes an extra pilot, navigator, or boom operator.

The crew is provided with diluter demand oxygen regulators adjacent to each crew seat within the cockpit. Additional diluter demand regulators are located within the boom pod. At altitude, when not at their assigned station, crew members carry emergency walk-around oxygen bottles.

A sextant port is located in the overhead of the cockpit. The sextant is generally shot by the boom operator.

Standard on all -135 series aircraft is the crew entry hatch and doorway. The hatch is hinged at the leading edge. A crew access ladder, normally stowed along the bottom of the main cargo door, is suspended from a receptacle located on the aft wall of the doorway. For safety purposes, a hinged grate is dropped over the doorway when personnel are not using the ladder. Should an emergency bailout be attempted, the crew entry door is jettisoned and a spoiler is dropped from the leading edge of the doorway. Few bailouts have been successful from any -135 aircraft.

The cabin sidewalls are lined with fold-down web troop seats. In addition, stanchions are provided for crew bunks. All passengers are issued emergency walk-around oxygen bottles.

Many airplanes have a dry galley attached to the cabin floor opposite the cargo door. This unit houses water and coffee inserts.

A basic flight lavatory is located forward of the cargo door on the left side of the aircraft. The lavatory includes a chemical toilet, two standing urinals, sink, fan, light, and waste receptacle. Some of the aircraft's electronics are also located within the lavatory. A sliding curtain mounted to a header bar provides privacy.

Risk kits are built up for deployments in which the required equipment is loaded on the cargo floor and strapped in place with web tiedown belts. Examples of risk kit equipment are tool boxes and spare landing gear/tire assemblies. During flight, the engine inlet and exhaust covers are stowed in the aft cabin. Crew and passenger luggage is loaded into large plywood bins which are erected on the cabin floor and strapped in place.

COLOR GALLERY

C-135B, 62-4127, was originally delivered to the 1611th ATW, McGuire AFB, New Jersey, on 31 May, 1962. By the time this photograph was taken, the aircraft had been converted to a VC-135B and assigned to the 89th MAW at Andrews AFB, Maryland. (Boeing-Wichita TW176-81C)

KC-135A, 57-2598, was assigned to CINCSAC. The aircraft was converted to a KC-135E in October 1989, and it was one of the first of these modifications. It is equipped with an HF antenna on the right wing tip, and it has a long-wire HF antenna.

KC-135A, 61-0316, was assigned to the U.S. Strike Command during the mid-1970s. Note how the white paint was applied to the wing tip and the leading edge of the wing. The overwing HF antenna was on the right wing only. The Strike Command emblem was located on the fin below the American flag. (via Remington)

KC-135 OFFICIAL PAINT SCHEMES

*Above: **Paso Del Norte** was the 300th KC-135A and was serial number 58-0085. As seen here at Moses Lake, Washington, it illustrates the original natural metal finish and high-visibility markings. A SAC band would later be applied to the Da-glo orange nose. (Williams)*

*Right: **Paso Del Norte** was assigned to Biggs AFB at El Paso, Texas, on 4 August, 1959. The 95th BMW insignia was later applied to the right side of the nose. (Williams)*

*During the early 1980s, KC-135As assigned to the 92nd BMW at Fairchild AFB, Washington, carried a light blue tail band with the name **Fairchild** in white script. Either a red or blue crown was applied below the tail band to indicate the 42nd or 92nd ARefS respectively. This Stratotanker illustrates the overall Aircraft Gray scheme.*

*The Baby Shamu camouflage scheme can be seen here on **PITTSBURGH'S FINEST** which is KC-135E, 59-1506. This aircraft is operated by the 171st ARefW, out of Pittsburgh, Pennsylvania. The yellow tail band with the state's name identifies the unit as does the shield on the nose. The aircraft has sixty-five mission markers from Operation Desert Shield that surround the blue map of Pennsylvania. This art is located just aft of the crew entry hatch.*

(171st ARefW, PAANG)

This KC-135R from the Mississippi ANG is painted in the new overall AMC gray scheme. It was photographed at Phoenix in January 1994.

COCKPIT & INTERIOR COLORS

A photograph of the pilot's station on a KC-135A from the 320th BMW reveals the original refueling panel. The factory-installed horn-buttons have long since been removed and replaced by home grown identifiers with the names of the unit and the base, thus thwarting the efforts of collectors.

This instrument panel was installed on a KC-135Q.

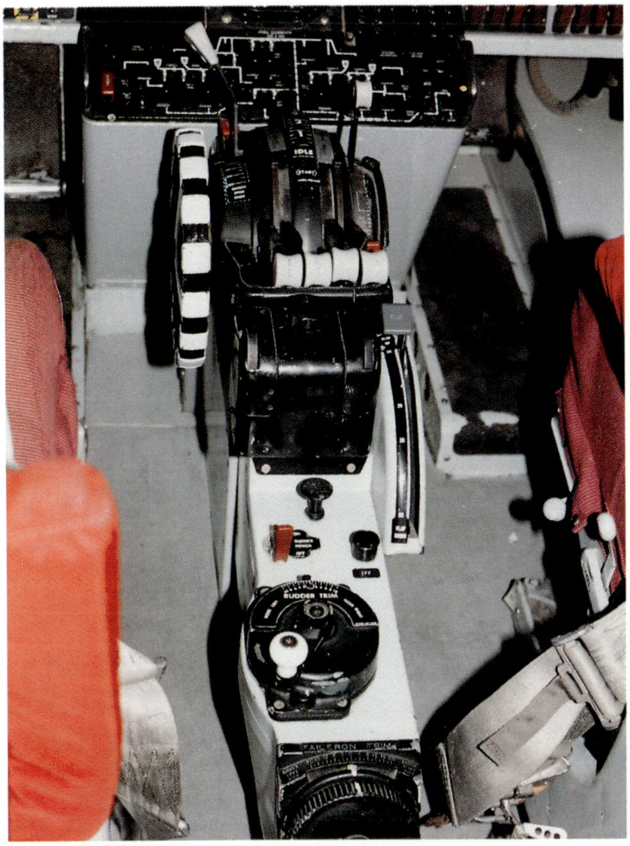

A typical aisle stand houses the throttle quadrant, speed brake lever, stabilizer trim wheel, stabilizer green band indicator, aileron and rudder trim wheels, and flap lever.

A circuit panel is located behind the boom operator's seat.

Details of the navigator's station can be seen here. If the pencil sharpener is whirring they're lost!

Here is a typical stowage box for walk-around oxygen bottles used by passengers. To the rear are four engine tail cone plugs, an APU, floor fan, and oxygen cylinders. Note the individualized walkway markings applied by the 940th ARefS, 320th BMW.

These three red ironing boards are located in the boom operator's pod. Note the chin rest at the center station.

NOSE ART OF THE 19th AIR REFUELING WING

*The 19th ARefW operates KC-135Rs from Robins AFB, Georgia. In October 1992, when the photographs on this page were taken, their tankers were painted in the Baby Shamu scheme, and most of them were adorned with nose art. This aircraft is 61-0307, and is named **WICKED WABBIT**.* *(Kinzey)*

*At left is a close-up of the nose art for **WICKED WABBIT** which is the aircraft shown above. At right is the art on 63-8002, which is named **Carpet Ride**.* *(Both Kinzey)*

***CHEROKEE ROSE** is 62-3554, while at right is **Lucky Strike** which is 60-0319.* *(Both Kinzey)*

***Jedi Knight** is shown at left, and is the nose art on 62-3530. **GLADIATOR** is painted on 57-2597.* *(Both Kinzey)*

BOEING KC-135 TANKER-TRANSPORT GENERAL ARRANGEMENT DRAWING

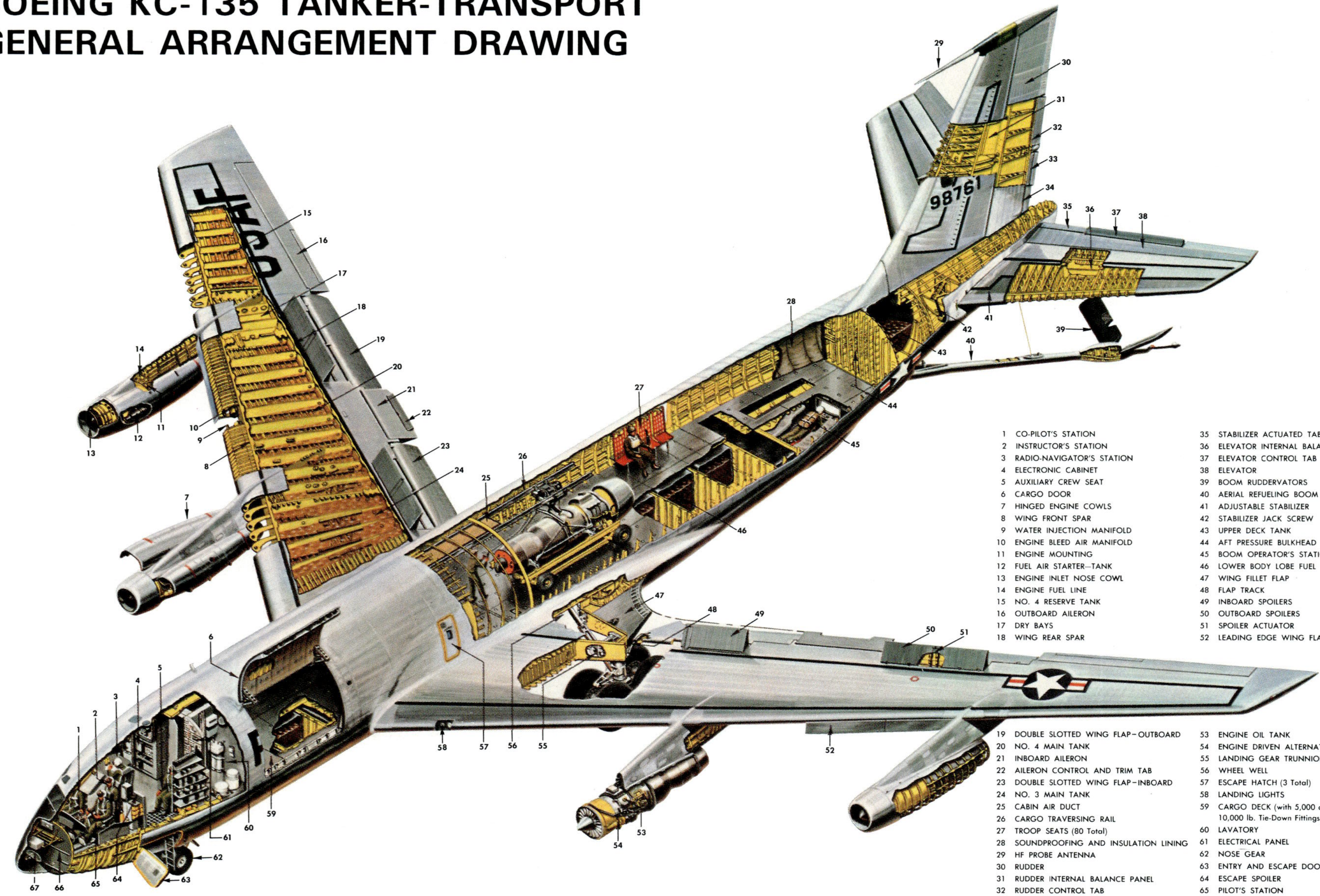

1 CO-PILOT'S STATION
2 INSTRUCTOR'S STATION
3 RADIO-NAVIGATOR'S STATION
4 ELECTRONIC CABINET
5 AUXILIARY CREW SEAT
6 CARGO DOOR
7 HINGED ENGINE COWLS
8 WING FRONT SPAR
9 WATER INJECTION MANIFOLD
10 ENGINE BLEED AIR MANIFOLD
11 ENGINE MOUNTING
12 FUEL AIR STARTER—TANK
13 ENGINE INLET NOSE COWL
14 ENGINE FUEL LINE
15 NO. 4 RESERVE TANK
16 OUTBOARD AILERON
17 DRY BAYS
18 WING REAR SPAR
19 DOUBLE SLOTTED WING FLAP—OUTBOARD
20 NO. 4 MAIN TANK
21 INBOARD AILERON
22 AILERON CONTROL AND TRIM TAB
23 DOUBLE SLOTTED WING FLAP—INBOARD
24 NO. 3 MAIN TANK
25 CABIN AIR DUCT
26 CARGO TRAVERSING RAIL
27 TROOP SEATS (80 Total)
28 SOUNDPROOFING AND INSULATION LINING
29 HF PROBE ANTENNA
30 RUDDER
31 RUDDER INTERNAL BALANCE PANEL
32 RUDDER CONTROL TAB
33 RUDDER ANTI-BALANCE TAB
34 RUDDER TRIM TAB
35 STABILIZER ACTUATED TAB
36 ELEVATOR INTERNAL BALANCE PANEL
37 ELEVATOR CONTROL TAB
38 ELEVATOR
39 BOOM RUDDERVATORS
40 AERIAL REFUELING BOOM
41 ADJUSTABLE STABILIZER
42 STABILIZER JACK SCREW
43 UPPER DECK TANK
44 AFT PRESSURE BULKHEAD
45 BOOM OPERATOR'S STATION
46 LOWER BODY LOBE FUEL CELL (9 Total)
47 WING FILLET FLAP
48 FLAP TRACK
49 INBOARD SPOILERS
50 OUTBOARD SPOILERS
51 SPOILER ACTUATOR
52 LEADING EDGE WING FLAP
53 ENGINE OIL TANK
54 ENGINE DRIVEN ALTERNATORS
55 LANDING GEAR TRUNNION
56 WHEEL WELL
57 ESCAPE HATCH (3 Total)
58 LANDING LIGHTS
59 CARGO DECK (with 5,000 and 10,000 lb. Tie-Down Fittings)
60 LAVATORY
61 ELECTRICAL PANEL
62 NOSE GEAR
63 ENTRY AND ESCAPE DOOR
64 ESCAPE SPOILER
65 PILOT'S STATION
66 FORWARD PRESSURE BULKHEAD
67 WEATHER RADAR SCANNER

C-135F

C-135F, 63-8471, was the first of twelve such aircraft to be delivered to the Armee d'la Air. *(Boeing P35133)*

A total of twelve C-135Fs were delivered to the Armee d'la Air in 1965. These aircraft were identical to the KC-135As with the exception that they had French roundels on the four wing positions and on the waist. However, U. S. Air Force serial numbers were carried on these airplanes.

These aircraft provide global capability to France's Dassault Mirage IV bombers, and they are equipped with the probe and drogue refueling system. C-135Fs were originally assigned to the 91st, 93rd, and 94th Escadre Bombardment (EB). As of 1 July, 1976, the 93rd Escadre was disbanded, and all C-135Fs were reassigned to the 91st and 94th Escadre de Revitailement en Vol [ERV] (Flight Refueling Squadrons). Squadron names and bases were:

UNIT	BASE	LOCATION
91st EB		
1/91st ERV	Gascogne	Mont de Marsan
4/91st ERV	Landes	Mont de Marsan
2/91st ERV	Bretagne	Cazaux
3/91st ERV	Beauvaisis	Criel
93rd EB		
1/93rd ERV	Guyenne	Istres
4/93rd ERV	Aunis	Istres
2/93rd ERV	Cevennes	Orange
3/93rd ERV	Sambre	Orange
94th EB		
1/94th ERV	Bourbonnais	Avord
4/94th ERV	Sologle	Avord
2/94th ERV	Marne	Saint-Dizier
3/94th ERV	Arbois	Saint-Dizier

As a result of the reorganization on 1 July, 1976, the units are now dispersed as follows:

91st EB		
1/91st ERV	Gascogne	Mont de Marsan
2/91st ERV	Bretagne	Cazaux
3/91st ERV	Beauvaisis	Criel
94th EB		
1/94th ERV	Bourbonnais	Avord
4/94th ERV	Sologle	Avord
2/94th ERV	Marne	Saint-Dizier

C-135Fs were known to have served operationally in Chad. Two of the aircraft from ERV 93 were deployed to N'Djamena on 21 August, 1968, where they supported Jaguars and Mirage F.1Cs. Several C-135Fs returned on 10 February, 1986, operating out of Bangui in support of Jaguars and Mirage F.1Cs as part of Operation Epervier (Sparrowhawk).

One C-135F, Serial Number 63-8473, was lost in the South Pacific during a French atomic weapons test on 1

July, 1972.

During the early 1980s, the remaining C-135Fs were returned to the Boeing plant in Wichita, Kansas, where high bypass GE/SNECMA F108-CF-100 (commercially known as CFM56-2) turbofans were retrofitted. These engines do not have thrust reversers. After modification, the aircraft were redesignated C-135FRs. As with the USAF KC-135Rs, these aircraft were retrofitted with a dual APU installation. Istres is the overhaul depot for all of the Armee d'la Air C-135FRs.

During Operation Desert Storm, several C-135FRs were dispatched to support the Armee d'la Air contingent which operated with the Coalition forces. The French called this Operation Daguet. A total of five C-135FRs were stationed at Al Ahsa, a civil airfield in eastern Saudi Arabia near the city of Al Hufuf. Between 2 August, 1990 and 16 January, 1991, the Armee d'la Air trained in Saddam Hussein's backyard during Operation Desert Shield.

The C-135FRs were deployed to the King Khalid International Airport near Riyadh for combat operations during Operation Desert Storm. The five aircraft were later joined by a sixth. Generally four aircraft were employed for the outbound portion of a mission and two for the retrurn. At the end of Operation Daguet, the C-135FRs had accrued 913 flight hours, while flying 327 sorties. They refueled:

185 Jaguars
437 Mirage 2000s
65 Mirage F.1CRs
8 F-14 Tomcats
1 EA-6B Prowler

Under a contract with France, Boeing is developing underwing refueling stores for the C-135FRs. The first kit was completed in mid-1993, and was followed by flight testing in Wichita. The remaining ten ship sets will be manufactured by October 1994.

Installation of the wingtip stores will permit simultaneous refueling of two fighters, while retaining the boom refueling capability for the French AWACSs aircraft which were delivered in 1991-1992.

C-135FR Serial Numbers

USAF	Armee d'la Air
63-8470	93-CA
63-8471	93-CB
63-8472	93-CC
63-8473	93-CD*
63-8474	93-CE
63-8475	93-CF
63-12735	93-CG
63-12736	93-CH
63-12737	93-CI
63-12738	93-CJ
63-12739	93-CK
63-12740	93-CL

* Not in Service

*C-135F, 63-12736, was the first in the series to be retrofitted with the GE/SNECMA high-bypass fan engines and given the designation C-135FR. The engine cowls and struts were painted Aircraft Gray. Note the code **CH**.*
(Boeing-Wichita 885-23)

C-135A & C-135E

This natural metal MATS C-135A is number 60-0369. It had Aircraft Gray paint on the engine struts, and the center sections of the cowls were white. The Da-glo high visibility colors of the day are also in evidence. (Boeing K8121)

Pending delivery of the Lockheed C-141 Starlifter, the USAF procured fifteen C-135A Stratolifters for use as interim transports. Even before the delivery of these aircraft, three boomless KC-135As, known as Falsies, were delivered to the Military Air Transport Service (MATS) until the C-135As were available. The first true MATS C-135A, 60-370, was delivered to McGuire AFB, New Jersey, on 12 August, 1961. While the Falsies had the early short KC-135A tail, the C-135As were delivered with the taller fin and boosted rudder.

The normal C-135A carried a crew of five. These included two pilots, a navigator, a flight engineer, and a loadmaster. The flight engineer occupied the jump seat between the pilots.

Powered by four Pratt & Whitney J57-P/F-59W turbojets, the C-135A had a maximum gross weight of 272,000 pounds and a range of 5,000 miles. While the refueling pod remained, the boom was replaced with a fuel dump tube. No upper deck fuel tanks were installed and the auxiliary power plant (APU) was deleted.

Other differences between the KC-135A and C-135A included the installation of a second air conditioning system, an electric versus manual emergency flap extension system, relocation of the emergency landing gear extension system from the main cabin to the cockpit, installation of miscellaneous passenger-associated equipment, and MATS-mission capable avionics. A hard shell cabin liner, with insulation, was added along with cabin lighting. Galley provisions were also installed.

The MATS avionics common to the C-135A included:

- Interphone system
- Public address system
- LORAN
- Three-light marker system (Outer, Middle, and Inner)
- Low range radar altimeter
- Dual low frequency ADF
- Dual VHF communications
- Dual HF communications
- Dual VHF navigation equipment
- Search radar
- High range radio altimeter
- IFF transponder
- Dual UHF communications
- Dual TACAN
- Dual glide slope system

The KC-135As were capable of carrying 126 troops or 89,000 pounds of cargo. Standard seating consisted of 126 troop seats in a six-abreast, aft facing configuration. In the cargo configuration, up to seventy-five seats could be installed. For aeromedical evacuation missions, a total of forty-four litter and fifty-four ambulatory patients could be carried. Two troop lavatories were located in the aft cabin. The crew used the standard lavatory in the forward cabin.

Thirteen of the C-135As were delivered to the 1611th Air Transport Wing (ATW) at McGuire AFB, one went to the 1501st ATW at Travis AFB, California, and one went to the Los Alamos Scientific Laboratories, Los Alamos, New Mexico. The last two aircraft will be discussed in second volume on the -135 in the Detail & Scale Series.

Subsequently, most of the C-135As were converted into a variety of unusual aircraft. Four C-135As were retrofitted with TF33-PW-102 engines and redesignated C-135Es. Two of these aircraft, 60-372 and 60-375, are assigned to Air Force Systems Command (AFCS), and the third, 60-376, is assigned to Air Combat Command (ACC), formerly known as the Tactical Air Command (TAC).

As of 1992, only two C-135As remained in the inventory. One was assigned to the Strategic Command (formerly SAC) as a command support aircraft, and the other was assigned to AFSC. These aircraft are 60-378 and 60-377 respectively.

C-135B, VC-135B, & C-135C

C-135B, 61-2663, is shown here as it appeared when it was delivered in a natural metal finish with Day-glo high visibility markings. The engine turbine cowls, wing lower surfaces (less the main gear doors), and elevators were painted Boeing 707 gray. This aircraft was delivered to Travis AFB, California, on 24 March, 1962. *(Boeing K8568)*

A total of thirty C-135B Stratolifters were produced for use by the Military Air Transport Service (MATS). The first of these aircraft flew on 15 February, 1962. Aircraft 61-2662 was the first to be delivered on 28 February, 1962.

The main difference between the C-135A and C-135B aircraft was the engines. Whereas the C-135As were powered by J57 turbojets, the C-135Bs had TF33-P-5 turbofans with thrust reversers. These engines gave the aircraft a maximum gross weight of 275,000 pounds. The aircraft featured a combination cartridge and pneumatic starter on the Number 3 engine. After starting the Number 3 engine with one of these systems, the remaining engines were started by using a cross-bleed starting technique. None of these aircraft were delivered with humped mounting struts for the engine nacelles.

Another difference was in the oxygen system. An integral emergency liquid oxygen system was installed in the C-135B, and this was a first in any military transports. If the cabin altitude exceeded 14,000 feet, the system automatically provided oxygen to the 126 troops for up to five hours.

MATS split the fleet equally between the 1611th Air Transport Wing (ATW) at McGuire AFB, New Jersey, and the 1501st ATW at Travis AFB, California. Crews immediately noticed the performance differences between these aircraft, particularly across the Pacific. Westbound C-135As always needed a refueling stop, whereas the C-135Bs could fly nonstop from Travis to Yokota AB, Japan. The installation of the fan engines increased the range from 5,000 to 8,000 statute miles. The C-135Bs could carry a nineteen-ton payload 5,500 miles from Yokota AB to Travis AFB, California, in less than nine and one-half hours.

C-135C, 61-2761, was photographed at Tinker AFB, Oklahoma. The aircraft was first delivered to Travis AFB on 30 April, 1962, as a C-135B. *(Ramsey)*

Also delivered to Travis AFB, C-135B, 61-2668, is shown here after conversion to a C-135C. It was one of two AMC aircraft assigned to support PACAF headquarters at Hickam AFB, Hawaii.

As of 1992 only four C-135Bs remained in the inventory, and all others have been converted for other missions. One C-135B, 62-4125, is assigned to U.S. Air Forces Europe (USAFE), while AMC operates 62-4126, 62-4127, and 62-4130. These aircraft were formerly designated VC-135Bs.

Three of the thirty C-135Bs delivered to MATS were modified to WC-135Bs in 1965. A pair of these were subsequently converted to staff transports in 1973, while the third aircraft was converted in 1974.

Three C-135As were retrofitted with TF33-PW-102 engines that were canabilized from commercial Model 707s. These aircraft were then redesignated as C-135Cs. The first two C-135Cs, 61-2668 and 61-2671, were assigned to MAC where they fly command support missions for Pacific Air Forces (PACAF). In 1975, the third aircraft, 61-2669, was converted into an avionics test bed for the Speckled Trout Program, and it was assigned to Air Force Systems Command, replacing KC-135A, 55-3126.

Two of the former WC-135Bs are currently operated by the new Air Mobility Command (AMC), and the third remains with AFSC. These aircraft were retrofitted with an air refueling receiver system when they were converted to the WC-135B configuration. This system remains on these aircraft even though they are no longer WC-135Bs.

In 1975, five MAC C-135Bs were upgraded for use as high level government transports and designated VC-135Bs. These aircraft were operated by the 89th Military Airlift Wing (MAC) at Andrews AFB, Maryland.

In addition to VIP interior accommodations, the aircraft were painted in the same distinctive medium blue, white, and gold scheme designed by First Lady Jacqueline Bouvier Kennedy for use on the Boeing VC-137Bs and VC-137Cs.

By August 1977, only two of these aircraft, serial numbers 62-4126 and 62-4127, remained with the 89th MAW. The three others reverted back to C-135Bs, with 62-4125 being reassigned to the 435th Tactical Airlift Wing (MAC) at Ramstein AB, Germany. There it was used as a VIP transport for USAFE. The other two were reassigned to the 55th SRW (SAC) for use as staff transports where they replaced three KC-135As.

In an effort to convince the American taxpayers that he was reducing the size of government, President Jimmy Carter downgraded the 89th MAW to a group and ordered that the two remaining VC-135Bs be converted back to C-135Bs. The distinctive blue, white, and gold scheme was replaced with a mundane overall white finish to give the impression of austerity. When the 89th was elevated back to wing status under President Ronald Reagan, the resplendant colors reappeared, however the VC-135B designation was never reinstituted. As of this writing, former VC-135Bs, 62-4126 and 64-4127, still serve with the 89th Wing of AMC as C-135Bs.

Originally delivered as a C-135B to Travis AFB on 29 May, 1962, 62-4125 was transferred to McGuire AFB in April 1964. This aircraft was photographed while assigned to the 89th MAW at Andrews AFB, and it is painted in the original VC-135B color scheme with the white engine cowls and struts.

RC-135A

RC-135A, 63-8058, was undergoing flight tests at Boeing Field when this photo was taken. Note the refueling pod and fuel dump tube. The flaps are in the landing configuration. These aircraft originally had MATS markings. (Boeing P38298)

Four RC-135As were delivered to MATS in 1965-1966. These aircraft had Fiscal Year 1963 serial numbers, but were delayed in delivery because of difficulties experienced in developing the mission-related equipment. The first two aircraft, 63-8059 and 63-8060, were delivered on 15 September, 1965.

The RC-135As were designed to supplant the RB-50s. With adequate fuel reserves, the RB-50s could fly approximately 3,800 miles at 20,000 feet; whereas the RC-135As could fly 4,900 nautical miles above 35,000 feet. This quadrupled the photographic coverage that could be accomplished on a single mission. The RC-135A was powered by four Pratt & Whitney J57-PW-59 water-injected turbojet engines.

The primary missions for these aircraft were photomapping and geodetic surveying, particularly for the down-range areas from Cape Canaveral. These aircraft were assigned to the 1370th Photo Mapping Wing, Air Photographic and Charting Service. APCS was part of the Military Air Transport Service (MATS) and was located at Turner AFB, Georgia. Prior to delivery of the aircraft, MATS was redesignated Military Airlift Command (MAC) and the markings on these aircraft were changed accordingly.

At the heart of the mission-oriented equipment was the Kollsman Instrument Corporation AN/ASQ-28 photomapping and geodetic surveying system. This system consisted of both air and ground equipment. The equipment carried in the aircraft was the AN/ASQ-92 subsystem which was electronically positioned. It permitted simultaneous, precise aerial photography, electronic and photographic terrain profile recording, and aerial geodetic surveying. The data gathered by the airborne subsystem permitted ground equipment to develop maps showing topographic features with controlled scale and perspective.

A camera compartment replaced a portion of the space usually occupied by the forward body fuel tank. Two optically ground camera windows, each thirty inches in diameter, were mounted in the bottom of this camera compartment. The glass in these windows was two and one-half inches thick. A door covered the two windows, and was slid aft on an external track when it was time to operate the cameras. Camera access was gained through a hatch in the cabin floor. The terrain profile recording antenna was installed in the former boom pod. Neither an air refueling system nor an air refueling receiver system was installed.

The RC-135A flight crew consisted of two pilots, a navigator, and a flight engineer. The mission crew was made up of a distance measuring interrogator, recording altimeter operator, and two photographers.

By July 1972, all four RC-135As were reassigned to SAC. While the camera compartment remained, all photographic equipment was removed and passenger seats were installed. These aircraft were employed for command support missions and served as staff transports. They subsequently were converted to the KC-135Ds.

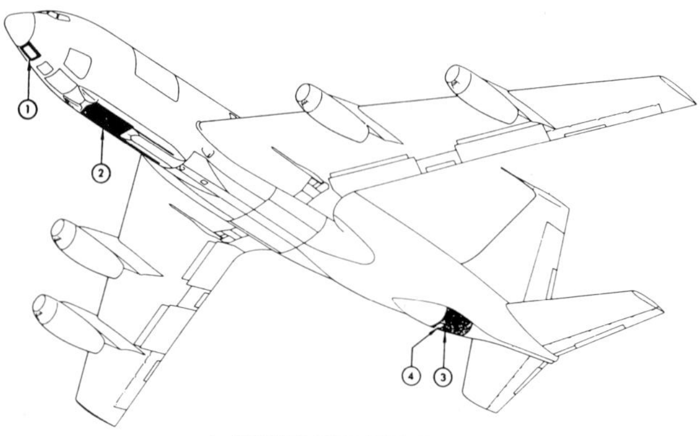

1. NAVIGATOR'S VIEW FINDER WINDOW
2. CAMERA WINDOW DOOR
3. TPR RADOME ACCESS
4. 35 MM CAMERA WINDOW

External features of the RC-135A are illustrated in this drawing.

When RC-135A, 63-8061, was delivered, MATS had already been redesignated as the Military Airlift Command (MAC), and the revised markings reflect this change. This was the last -135 aircraft to be delivered, and it was turned over to the Air Force on 1 June, 1966. *(Boeing P39951)*

Details of the camera bay door and revised refueling pod are visible in this view. *(Boeing P39956)*

KC-135D

All four RC-135As, including 63-8060, were transferred to SAC for use as command support aircraft. A white cap and blue cheat line were applied to these aircraft.
(USAF)

When SAC obtained the four RC-135As from MAC, they were used as staff transports and were assigned to Offutt AFB. Later, when SAC needed additional tankers, these aircraft were converted to serve in the standard tanker role during late 1980. An air refueling boom, boom pod station, refueling director lights, and associated equipment were retrofitted. A number of minor variances in such systems as anti-skid brakes, communications equipment, and the air conditioning system resulted in these tankers being designated KC- 135Ds. All four of these tanker aircraft were initially assigned to the 305th ARefW at Grissom AFB, Indiana.

In 1990, these four aircraft were retrofitted with TF33-PW-102 turbofan engines instead of the factory-installed J57s. Because these aircraft were unique anyway, they retained their KC-135D designation. After the engine retrofit, all four aircraft were assigned to the newly formed 168th ARefG of the Alaska ANG, and they were based at Eielson AFB, Alaska.

Reconfigured as a KC-135D, 63-8053 was photographed lifting off from Greenville, Texas. A refueling system had been retrofitted to the aircraft.
(E-Systems P-16539-2)

KC-135E

KC-135E, 58-0080, is assigned to the 151st ARefG of the Royal Mormon Air Force, also known as the Utah ANG. They are noted for displaying their unit's flag through the sextant port while taxiing into other bases.
(Boeing-Wichita 1190-1141)

In 1979, while the Air Force was pursuing the CFM56 high bypass turbofan engine for upgrading SAC's tanker fleet, a less expensive program was conceived for the Air National Guard (ANG) and Air Force Reserve (AFRes) tankers.

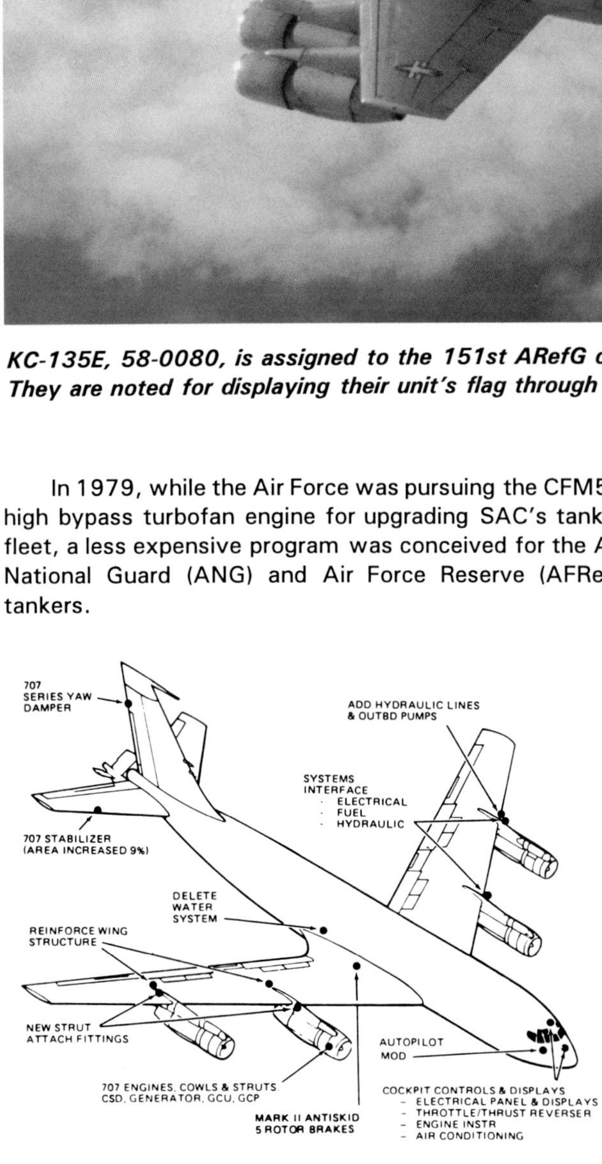

Retrofitted features of the KC-135E are indicated in this drawing.

The first KC-135A to be converted to the KC-135E configuration was serial number 59-1415. The aircraft entered the modification line at Boeing Wichita on 30 September, 1981, and made its innaugural flight on 19 January, 1982. It was formally presented to the USAF during a ceremony on 26 January, 1982.

During the first seven months of 1982, a total of eighteen special purpose -135s had their J57 engines replaced with the TF33-PW-102 powerplants. These were followed by the KC-135s assigned to the Air National Guard and Air Force Reserve.

The Air National Guard and Reserve KC-135As were retrofitted with the JT3D-3B engines removed from commercial Model 707 airliners. Basically this engine is the same as the Air Force's Pratt & Whitney TF33-PW-5s, but it has one additional compressor stage. The commercial JT3D has the USAF designation TF33-PW-102. This retrofit offered significant state-of-the-art improvements over the earlier 1950's vintage J57s. These engines retained their commercial thrust reversers, thus enhancing wet runway operations. KC-135As retrofitted with these engines are redesignated KC-135Es (not to be confused with the RC-135E). An almost twenty percent increase in fuel offload capability is realized because of this modification. Since the engines came from a variety of commercial Model 707s, the humped struts appear in various positions on different aircraft. Some KC-135Es have three

The blue trim on the AFRES fin flash indicates that KC-135E, 58-0041, was operated by the 72nd ARefS, 931st ARefG.

humped engine struts while others have two.

The higher thrust of the turbofan engines also necessitated the installation of horizontal stabilizers with greater surface area to assure pitch control. The simple solution was to utilize the stabilizers from the cannibalized commercial jetliners. This was done on the first eighty-five aircraft only. Later conversions used locally fabricated tip extensions on the existing stabilizers.

The series yaw dampers used on the Model 707s were also removed and installed on the KC-135Es. The series yaw damper operates full time to improve lateral and directional airplane handling qualities (i.e. Dutch roll). The system is designed to operate without feedback through the pilots' rudder pedals.

Many Guard and Reserve units have gone beyond the basic amenities authorized by the Air Force and have provided their own unique creature comforts. Some aircraft have even been outfitted with airline-type passenger seats. In some configurations the seats face each other and are separated by a table. To show pride in their unit and personnel, many of these aircraft feature reader boards on the aft face of the lavatory. A wide variety of carved and lighted signs identify the unit, aircraft name, crew chief, and assistant crew chiefs.

With the exception of a single KC-135E which was assigned to SAC headquarters, all of this series are operated by the Air National Guard and Air Force Reserve tanker units. These units are as follows:

Air National Guard

101st ARefW	Bangor, Maine
108th ARefW	McGuire AFB, New Jersey
126th ARefW	Chicago, Illinois
141st ARefW	Fairchild AFB, Washington
171st ARefW	Pittsburgh, Pennsylvania
112th ARefG	Pittsburgh, Pennsylvania
128th ARefG*	Milwaukee, Wisconsin
134th ARefG	Knoxville, Tennessee
151st ARefG	Salt Lake City, Utah
157th ARefG	Portsmouth. New Hampshire
160th ARefG*	Rickenbacker ANGB, Ohio
161st ARefG	Phoenix, Arizona
168th ARefG	Eielson AFB, Arkansas
170th ARefG	McGuire AFB, New Jersey
190th ARefG	Topeka, Kansas

* These units are now equipped with KC-135Rs.

Air Force Reserve

72nd ARefS, 434th ARefG, Grissom AFB, Indiana
314th ARefS, 452nd ARefG, March AFB, California
336th ARefS, 452nd ARefG, March AFB, California

KC-135Q

KC-135Q, 58-0129, was flown by the 100th SRW at Beale AFB, California. The tanker was refueling SR-71, 61-7978, when this photo was taken. (Dunham)

Fifty-six KC-135As were modified to KC-135Qs for SAC to support their Lockheed SR-71 Blackbirds. These aircraft have the same external appearance as the KC-135A, except that they had an additional TACAN antenna which was used to rendezvous with the SR-71s. Internally, the KC-135Q has a modified fuel system to accommodate the JP-7 fuel that was used by the SR-71s.

Most of the KC-135Qs were stationed at Beale AFB, California, and operated first by the 100th SRW, and then by the 9th SRW after the 100th was inactivated. In addition, a number of KC-135Qs were assigned to other tanker units in order to provide worldwide support of SR-71s. Two squadrons at Beale are still equipped with these aircraft, even though the SR-71s have been retired.

As an interim change, the KC-135Qs were retrofitted with the TF33-PW-102 turbofans and cannibalized horizontal stabilizers. Because of the uniqueness of these aircraft, there was no change in aircraft designation to reflect this engine change. When funding becomes available, the KC-135Qs are scheduled to be refitted with the CFM56-2 high-bypass turbofan engines. If and when this occurs, their designation will be changed to KC-135T.

KC-135Q, 60-0346, has the last SAC-generated camouflage paint scheme, and it retains its fin flash. As of January 1993, this aircraft had yet to be retrofitted with turbofan engines. Now part of ACC, this aircraft has the ACC crest on the fin, and a fighter-style tail code and number. The BB tail code is indicative of the unit's nickname, the Beale Bandits. (Doty)

KC-135R

KC-135A, 62-3556, tests the air refueling receiver system for the prototype KC-135R which was serial number 61-0293. Note the instrument probe. ***(Boeing-Wichita 882-617)***

High-bypass General Electric/SNECMA F108-CF-100 (commercially known as CFM56-2) engines were retrofitted to some KC-135As beginning in 1984. These engines do not have thrust reversers. While reversers are standard on commercial CFM56 engines, the Air Force believed that the maintenance costs would be higher, so it paid a higher price to have them deleted during production. However, consultation with a few airlines would have shown that higher maintenance costs were not the case. As a result of this decision, many Air National Guard KC-135Es, which do have thrust reversers, are deployed to the northern United States during the winter because of their better wet runway operating characteristics. Airplanes for SAC are redesignated as KC-135Rs (not to be confused with the 1960s-vintage KC-135Rs which will be covered in Volume 2). A similar modification to the French C-135Fs resulted in C-135FRs.

The KC-135R is more fuel efficient and this retrofit results in a fifty-percent greater fuel offload than possible with the KC-135A. Installation of the F-108-CF-100 engines also necessitated the installation of larger horizontal stabilizers. Other modifications were made to the flight

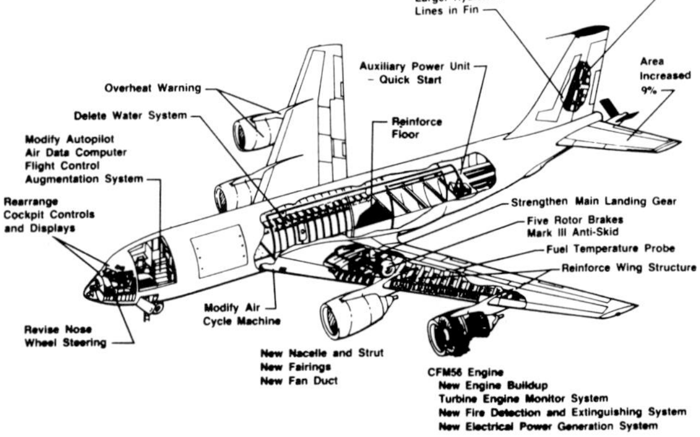

Retrofitted features of the KC-135R are indicated in this drawing.

KC-135R, 62-3511, is from the 384th BMW, and it displays its post-1987 fin flash. McConnell is in black above a medium blue band which is bordered in black.

KC-135R, 59-1455, passes Mount Rushmore. The features of this historic landmark are incorporated into the fin flash. ***(USAF)***

control system, electrical power generation system, engine fire/overheat detection and extinguishing system, anti-skid system, and the load-bearing capability of the landing gear.

Of interest is the flight control augmentation system (FCAS) which provides automatic rudder control imputs to assist the pilots with yaw control in the event of an outboard engine failure. Also, a series yaw damper, as employed on commercial Boeing jetliners, was retrofitted to provide automatic rudder inputs to respond to a broad range of adverse yaw conditions experienced by swept-wing aircraft.

On 22 June, 1982, the first KC-135R, 61-0293, was rolled out of the Boeing Wichita plant and made its first flight on 4 August. After completing flight testing, the aircraft was delivered directly across the runway to the 384th Air Refueling Wing at McConnell AFB, Kansas, on 20 June, 1984.

The KC-135Rs were initially assigned only to active duty Air Force refueling units. Beginning in the spring of 1992, several ANG units were equipped with the KC-135R. These units are:

128th ARefG	Milwaukee, Wisconsin
160th ARefG	Rickenbacker ANGB, Ohio
186th ARefG	Meridian, Mississippi.

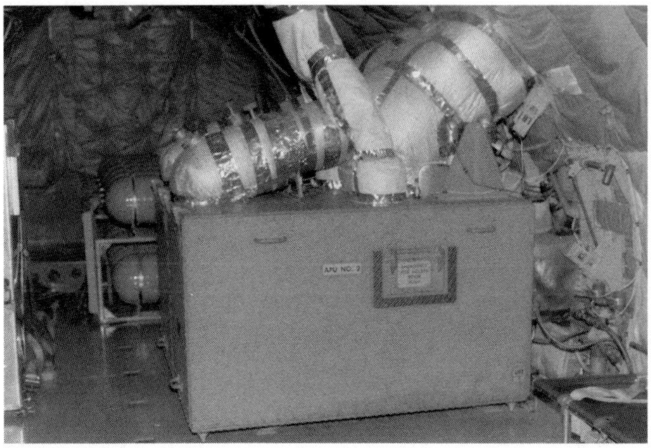

This auxiliary power unit (APU) was installed in the aft fuselage of KC-135Rs.

Details of the APU inlets and exhausts on the KC-135R are revealed here. Note the reinforcing doublers.

PAINT SCHEMES & MARKINGS

The original natural metal finish and early markings can be seen on these two aircraft. KC-135A, 58-0084, is in the foreground, and the 11th BMW insignia can be seen on SAC's Milky Way band. This band was originally placed in this position on the nose and was located on the Day-glo orange band as seen here. The SAC insignia was in the same place on the opposite side. In the background is KC-135A, 58-0035, which was assigned to the 7th BMW. This photograph dates from the Strategic Air Command's 1959 Bomb Competition. (USAF)

This section outlines the basic paint schemes, colors, and markings applied to various KC-135 and C-135 aircraft. It also explains the important changes that were made to the schemes and markings over the many years these aircraft have remained in service.

Basic Finishes and Paint Schemes

Originally, all C-135 aircraft were delivered in a natural metal finish, except that magnesium parts were painted light gray. Wheel wells and flap wells were painted chromate (yellow/green), and the landing gear was a silver color officially called Aluminum. High-visibility markings were applied to the nose, tail band, and wing tips, this color officially being yellow/orange Da-Glo, AN 634 (FS 28915). The Da-Glo was later deleted because of corrosion problems. Also, the orange faded within about six months to a dirty white. This resulted in a blotchy appearance with looked to be almost all white on the upper surfaces, but which graduated to an almost new-looking Da-Glo on the lower surfaces of the aircraft. Subsequently the -135s were delivered without Da-Glo paint, and the overall airframe was painted with a metallic-colored paint that was officially named Gloss Silver (FS 17178). Modelers should note that Testors Metallic Silver is a reasonable approximation of this silver

Early MATS markings can be seen on these two C-135Bs as they are checked over in the fuel dock at Renton Field. Note the high-visibility, Day-glo orange band around the aft fuselage. The MATS insignia is located within this band. ***(USAF)***

color. The silver anti-erosion finish that was applied to all leading edges was 3M Corogard. This can be represented on models with Pactra Candy Undercoat Silver.

During the mid-1970s an overall gray scheme appeared as specified in MIL-C-83286. This alaphatic polyurethane paint was gloss Aircraft Gray (FS 16473). The radomes were gloss black (FS 17038), while the anti-glare panel was flat black (FS 37038). A chemical and solvent resistant finish of either gloss white (FS 17875) or gloss Aircraft Gray was applied to the engine cowls. The landing gear struts were painted gloss Insignia White (FS 17875), and the wheel wells were painted gloss Aircraft Gray.

During the mid-to-late 1980s, SAC conceived low-level operations for the tankers, and the upper surfaces were oversprayed Dark Gray (FS 36081). According to Dana Bell, the noted aviation researcher and author, this color, which was employed in TAC's European I scheme and which was the darker gray in SAC's Strategic Camouflage Scheme, tends to fade towards a dark green. This scheme was nicknamed Baby Shamu when used on KC-135s, but was simply called Shamu when applied to the larger KC-10s. SAC had planned to replace the FS 36081 gray with an overall Dark Ghost Gray (FS 36320). However, no known use of this color has been found.

In early 1992, Air Force Chief of Staff, General Merrill A. "Tony" McPeak, announced some startling reorganization plans for the U. S. Air Force. It was a sad day for the late General Curtis E. LeMay, father of the Strategic Air Command. The three major flying commands, SAC, TAC, and MAC were replaced by a new Air Combat Command (ACC), responsible for all shooters and reconnaissance operations, and the Air Mobility Command (AMC) which gained almost all of the transport and tanker assets. This reorganization resulted in still another paint scheme for the KC-135s. The new Air Mobility Command is scheduled to have its KC-135s repainted in a new gray color as they go through their regularly schedueled depot maintenance. This new gray is officially known as flat Battle Gray (FS 36173).

Basic Official U. S. Air Force Markings

On uncamouflaged aircraft, **U. S. AIR FORCE** appears in thirty-inch gloss Insignia Blue (FS 15044) letters on both sides of the fuselage beginning with Body Station 257.00. A national insigna with a fifty-inch diameter is placed on the aft fuselage and is centered at Body Station 1310.00. Insignias of the same size are also on the wings parallel with the rear spar and thirty-one inches out-

The overall Aircraft Gray scheme is illustrated on KC-135A, 61-0305, which was assigned to the 305th ARefW at Grissom AFB, Indiana, when this photograph was taken. Note that the Milky Way band has been moved to the center fuselage. These are post-1987markings for the 305th ARefW, and they include a black and white checkered fin flash and the Pacesetters racing flag emblem to represent the Indianapolis 500.

The Baby Shamu camouflage scheme is shown here on KC-135R, 57-1427. This was the last paint scheme generated by the Strategic Air Command, and it was also used on Air National Guard and Air Force Reserve aircraft. This particular tanker is from the 160th ARefG of the Ohio Air National Guard. *(160th ARefG, OHANG)*

board of the outer engine strut. **USAF** is stencilled in gloss Insignia Blue letters that are fifty-five inches high and which are located on the top right and lower left outboard wing panels. These markings are centered on the rear spar and located so that the lower inboard corner is thirty-one inches outboard of the centerline of the outer engine strut. Radio call numbers are painted in eighteen-inch gloss black (FS 17083) numerals and are located on both sides of the vertical stabilizer. A two-inch wide gloss black walkway marking appears only on the left side of the folding vertical fin. A two-inch gloss black walkway is painted along the spars on the upper surfaces of the wings and horizontal stabilizers.

On camouflaged KC-135s, a nine-inch **USAF** and radio call number appear in black on the vertical stabilizer. The national insiginas are black stencils and are eighteen inches in diameter. They appear along the aircraft centerline at Body Station 1310.00 and on the top left and bottom right outboard wing panel. These insignia are tangent to the wing rear spar. The gloss black radome is retained as are the walkway markings.

A pair of two-inch wide red warning stripes were placed in the compressor and turbine areas for many years. However, these warning stripes have long since been removed. With the introduction of the high bypass F108-CF-100 engines, concern arose about the safety of ground personnel. Boeing introduced a set of engine cowl warning markings for the commercial CFM56-3 installation on the later Model 737 series airplanes, and these warning markings were adopted by the Air Force.

A two-inch wide Orange-Yellow (FS 13538) marking is applied around the forward entry way and emergency exit hatches on the aircraft that are painted Aircraft Gray. These same markings are red on the Baby Shamu scheme.

The receiver alignment stripe, which is sixty-five feet long and eight inches wide, is painted on the belly. It was originally a fluorescent yellow Scotchcal, but now this stripe is painted yellow.

The boom ruddevators are painted flat black. They once carried white codes such as AH10, JJ32, and LL32 for radio silent operations conducted under Emergency War Order training exercises. These boom codes are no longer used, however, Air National Guard units proudly display their unit identifier on the ruddevators.

SAC Markings

Originally KC-135s assigned to SAC carried the Milky Way band on the nose. These bands were subsequently moved to the waist. The SAC crest was applied to the left while a wing insignia was located on the right, each being centered on the Milky Way band. When the Baby Shamu scheme was introduced, the Milky Way band was deleted. Many units carried a colored identifying band at the top of the vertical tail. Beginning in the mid-1980s SAC authorized a series of fin tip markings for each unit.

Unit Markings

Traditionally, the SAC unit identification consisted of the wing crest being centered on the Milky Way band on the right side, and occasionally a colored tail stripe was used. When the Baby Shamu scheme was introduced, the low visibility SAC and unit crests were applied to the right and left sides of the nose respectively. When the Air Mobility Command gained SAC assets during the reorganization of the Air Force, the AMC crest replaced the SAC insignia.

During the mid-1980s SAC authorized a series of geometric patterns that were applied to the upper fins for unit identification. These short-lived markings were again changed during the late 1980s when the units were allowed to use stylized markings on the fin tips. These markings were in full color on the overall Aircraft Gray tankers and in subdued colors on the camouflaged aircraft.

Air National Guard Markings

In 1975, when the first KC-135As were assigned to the Air National Guard, their markings were identical to those on SAC aircraft with one addition. The name of the state was lettered in black above the ANG crest on the vertical fin. During America's bicentennial celebration in 1976, a commemorative ANG crest was applied. The SAC Milky Way band was part of the ANG markings, and the unit crest was applied to the band on the right side. During the early 1980s, the state name was applied within a colored band on the top of the vertical fin, and the aircraft were painted overall Aircraft Gray. When SAC introduced the Baby Shamu scheme, many ANG aircraft received the same treatment as SAC aircraft when they went through depot maintenance.

Air Force Reserve Markings

Also in 1975, three Air Force Reserve units transitioned to the KC-135A, and their markings were identical to SAC's. However, the Air Force Reserve unit crest replaced the SAC wing crest on the right side of the Milky Way band. The vertical fin carried a broad white band with three parallel stripes and an arrowhead at the leading edge. **AFRES** was lettered on the tail in unit colors.

1992-Era Basic Markings - AMC

Beginning in 1991, a new AMC directive called for the aircraft to be painted an overall flat Battle Gray (FS 36173). The two-inch wide gloss black walkway markings have been deleted from the wings and empennage.

A pair of two-inch wide flat black stripes are applied to the top of the vertical fin. One stripe is located at the base of the fin tip antenna fairing and the second is twelve inches below it. The stripes are parallel to the waterlines and start aft of the fin's leading edge panel line. Currently these stripes are not authorized for Guard -135s, because there is a marking identifying the state to which the aircraft is assigned in this location.

A new matte U.S. flag decal, which measures twenty-one by forty inches, is applied to the vertical stabilizer with the blue field forward and the upper forward corner tangent to the fin's leading edge panel line. The bottom

KC-135A, 58-0088, displays the new 1992-era overall gray paint scheme, and it has markings from the 2nd Wing based at Barksdale AFB, Louisiana. The ACC tail code and fighter-style tail number are in evidence. Barksdale is lettered in black within the medium blue fin flash.

edge is approximately forty-four inches above the UHF antenna. A major command (MAJCOM) identifer, **AMC**, is painted in twelve-inch flat black letters and is centered twelve inches below the flag. The five-digit radio call number is applied in twelve-inch flat black numerals, and it is centered twelve inches below the MAJCOM identifier.

1992-Era Basic Markings - ACC

Tail markings for the Air Combat Command are different from those used by AMC. The following information concerning these markings applies to all ACC -135 aircraft with two exceptions. These exceptions are KC- 135E, 57-2589, which is assigned to the STRATCOM Commander-In-Chief, and KC-135E, 59-1514, assigned to the Boeing Military Airplane Company in Wichita, Kansas, for test purposes.

The depot applies the tail number in black with **AF** and the last two digits of the contract year in eight-inch characters. The last three digits of the serial number are painted in eighteen-inch numerals.

Field units apply a fifteen-inch wide fin flash and the ACC insignia. In addition, a two-letter tail code for the assigned base is painted in flat black letters on the fin. These letters are thirty-six inches high.

Effective 1 October, 1993, all Air Combat Command KC-135A/E/Q/R/T tankers (except for six KC-135Rs with the 366th Wing, Mountian Home AFB, Idaho, and two KC-135Es with the 55th SRW, Offutt AFB, Nebraska) were transferred to the Air Mobility Command.

1991-Era Basic Markings - PACAF

The depot applies the tail number in black with **AF** and the last two digits of the contract year in characters which are six-and-one-half inches high. The last three digits of the serial number are numerals which are fifteen-inches in height.

Field units apply the fifteen-inch flat black unit tail code above the tail number. These units are also responsible for adding the PACAF insignia, which is twenty-four inches high and a fin flash that is fifteen inches wide.

1992-Era Basic Markings - USAFE

The depot applies a twelve-inch flat black radio call number below the UHF antenna, and a twelve-inch flat black **USAF** two inches above the antenna.

The field unit adds a red, white, and blue fin flash with nine white stars, and the European Tanker Task Force (ETTF) crown and insignia at the top of the fin. In addition, a square **D** (based on the old 100th BG marking used in World War II) is applied between the fin flash and **USAF**.

1992-ERA Basic Markings - Common

While the two-inch wide walkway markings have been deleted, seven **NO STEP FWD** and eight **NO STEP AFT** markings are applied along the front and rear spars of the wings in flat black letters that measure one inch in height.

The exit hatch outline markings are applied in the form of flat black stripes with rounded corners. These stripes are one inch wide. The crew entry door has a surrounding stripe which is one-half inch wide and a pair of hatch latch pushbutton markings. **OPEN LATCHES-PULL HANDLE DOWN** is lettered in flat black.

The radio call number is applied to the left and right sides of the forward fuselage in six-inch flat black letters.

1992-Era Tail Codes

BB	9th Wing	Beale AFB, California
DY	97th Wing	Dyess AFB, Texas
LA	2nd Wing	Barksdale AFB, Louisiana
MO	366th Wing	Mountain Home AFB, Idaho
OF	55th Wing	Offutt, AFB, Nebraska
ZZ	909th ARefS	(PACAF) Kadena AB, Okinawa

KC-135A, 63-8023, was serving with the 376th SW at Kadena AB, Okinawa, when this photograph was taken. Above the crew entry door is a USAF Outstanding Unit Award. To commemorate the aircraft's visits to Diego Garcia, it was zapped with Diego lobsters. (USAF)

Zaps

Zaps have been applied to the noses of many aircraft when they transited other bases. Examples include lobsters from Maine and Diego Garcia, kangaroos from Australia, and silhouettes of Wake Island for visits to that Pacific island.

Mission Markings

During Operations Desert Shield and Desert Storm, participating units developed different styles of mission markings. During Desert Shield, an oil derrick or camels were often applied, while during Desert Storm lightning bolts or eagles usually represented the missions flown.

These markings on 58-0008 are typical of those applied to KC-135Es from the 160th ARefG during deployment for the Gulf War. There are two symbols above the crew entry hatch. The one on the left is green and is for Desert Shield, while the one on the right is red and symbolizes the aircraft's participation in Desert Storm. The red camels below the nose number are for missions flown during Desert Shield, and the flying egale markings aft of the entry hatch are for missions flown during Operation Desert Storm. On may KC-135s, lightning bolts were used to represent the missions flown during Desert Storm. (160th ARefG, Ohio ANG)

NOSE ART

***Back Off** is KC-135E, 56-3607, which was assigned to the 134th ARefG. It carries Arabic characters and nine mission symbols from Operation Desert Shield.*

***Classic Hog** is the name of KC-135E, 59-1485, of the 170th AREfG. It also served in the Gulf War.*

Nose Art

During the late-1980s, CINCSAC, General John T. "Jack" Chain authorized the return of nose art on the bombers and tankers. As a result, a plethora of revitalized World War II markings and completely new artwork appeared.

With the reorganization of the U. S. Air Force, and in the wake of the Navy's Tailhook scandal, a new era of personalized markings has begun. On 24 July, 1992, the new Air Mobility Command vice commander issued a directive to the Fifteenth Air Force (new USAF tanker organization) providing a new set of guidelines for nose art. It stated, in part, "...we have had many discussions on how to ensure this practice is consistant with our broader mobility mission and the numerous sensitivities of people of virtually every sex, race, and religious belief our aircraft will represent or be displayed before as they perform our worldwide mission. This guidance will be formalized in AMCR 66-25, soon to be published."

Basically, nose art can be continued if it reflects a theme of civic and community pride, be distinctive, symbolic, and be designed and maintained to the highest quality standards. Designs will be gender neutral and reviewed and approved by the wing commander prior to installation. A color copy will be forwarded to Headquarters AMC and AFMC. Nose art on KC-135s will be in the left and/or right forward fuselage, not to exceed an area four feet high by three feet wide.

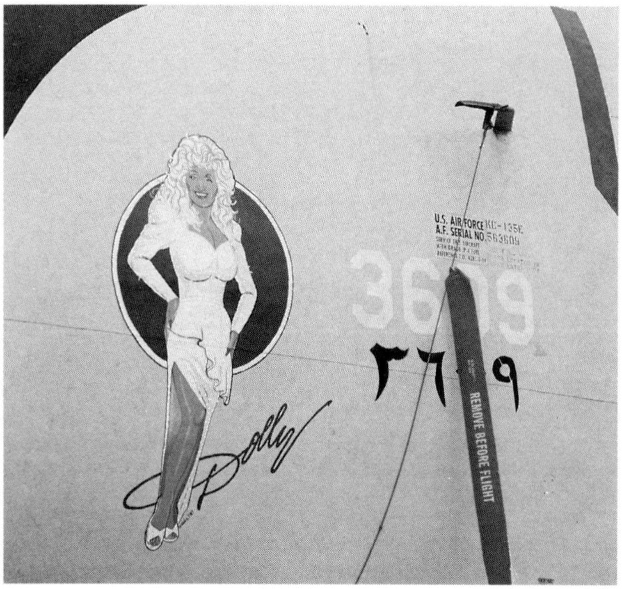

***Dolly** is KC-135E, 56-3609, from the 170th ARefG. This aircraft is another Gulf War veteran.*

***Aces High** is KC-135E, 58-0078, which is assigned to the 170th ARefG. Note the Arabic characters remaining from the Gulf War. The Aces theme is in memory of the flying aces of WW I and WW II, while the Jersey Devil is in reference to the state's legalized gambling.*

(170th ARefG, NJANG)

Easy Money *is KC-135E, 58-0040, which is operated by the 170th ArefG. It is stationed at McGuire AFB, New Jersey. The art is based on the Rodney Dangerfield movie of the same name, and the girl in a flight suit is superimposed on a map of New Jersey. (170th ARefG, NJANG)*

Bottoms Up *is KC-135E, 59-1497, which is flown by the 170th ARefG. The theme ties the 170th together with its sister unit, the 108th ARefW. The 108th was also based at McGuire AFB, and it previously flew F-4 Phantoms. (170th ARefG, NJANG)*

Northern Exposure, *which was named for the television series, is KC-135R, 62-3498. It is operated by the 453rd ARefG.*

Wind Rider *was KC-135A, 61-0267, from the 97th BMW. The aircraft was converted to a KC-135E in 1990. (Thompson)*

Fuel Mule *is KC-135R, 60-0328, and it is assigned to the 453rd ARefG at Fairchild AFB, Washington. Note that the Seahawk fin flash remains with these new Air Mobility Command markings.*

City of Blytheville *was KC-135A, 63-8012, which was flown by the 97th BMW. When the picture was taken, the pilot was LTC John Horner, commander of the 97th ARefS. (Thompson)*

NOSES & TAILS

*The three photographs above show the markings on KC-135A, 57-1432, when it was assigned to the 92nd BMW and based at Fairchild AFB, Washington. After the aircraft had experienced a rash of boom problems, it was named **Eeyore**. Artwork, which is shown in the top left photograph, was painted on the right side of the nose instead of the left side where such art is usually found. However, the name was also lettered in the top stripe of the Seahawk emblem on the left side of the nose. The only unit marking on the vertical tail was the Seahawk emblem which was positioned just above the rudder. Modelers will recognize these markings as the ones provided in ERTL's 1/72nd scale KC-135 kit.*

*Sometimes the art is on both sides of the nose as is the case of **SOARING EAGLE**. This aircraft is KC-135R, 61-0295, which is assigned to the 340th ARefW at Altus AFB, Oklahoma. The fin flash consists of a medium blue Conestoga wagon within a map of Oklahoma. A small white star in the lower left corner of the map represents the location of the base.*

KC-135s are sometimes named but do not have elaborate nose art. In such cases, the names are usually lettered within the crew chief's block. ***BETTY*** *was the name given to KC-135R, 60-0366, when it was assigned to the 301st ARefW at Malmstrom AFB, Montana.* ***BETTY*** *is in very small letters and is the topmost lettering within the silhouette of Montana. The wing insignia was applied to the left side of the nose, while the SAC crest appeared on the right. Note that the flat black anti-glare panel was retained on the camouflaged aircraft. At right is a close-up of the tail flash on the same aircraft. These were the markings used by the 301st ARefW during 1991.*

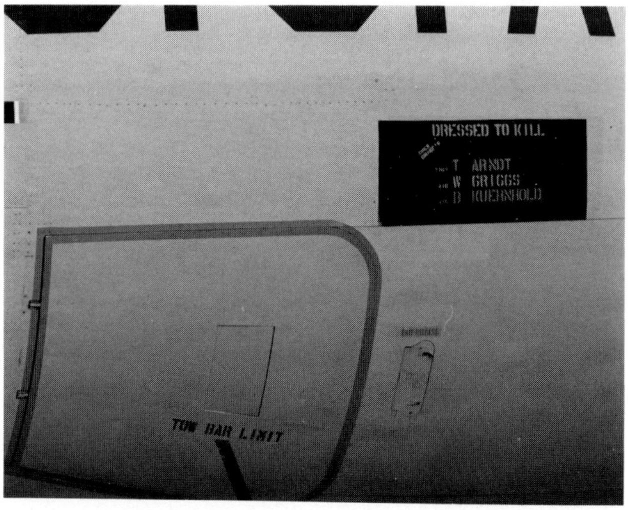

Left and above: ***Dressed to Kill*** *was the name given to KC-135A, 62-3560, while it was assigned to the 5th BMW. The aircraft was deployed to Eielson AFB, Alaska, during the summer of 1991 where it served with the Alaskan Tanker Task Force. The fin flash consisted of a white band, an orange sun ball, and a black aircraft silhouette. The 5th BMW death's head appeared on the fin, and the full-color wing crest was applied to the Milky Way band. The aircraft's name appeared in the crew chief's block.*

USAF TAIL MARKINGS

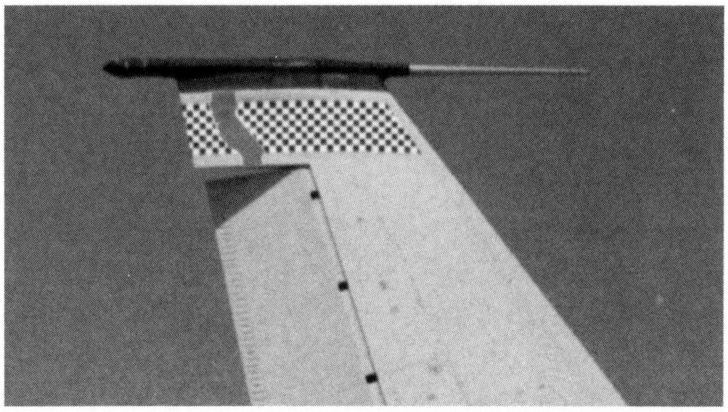

The 940th AResS of the 320th BMW uses a black and white fin flash. A yellow map of California is located within the flash above the rudder.

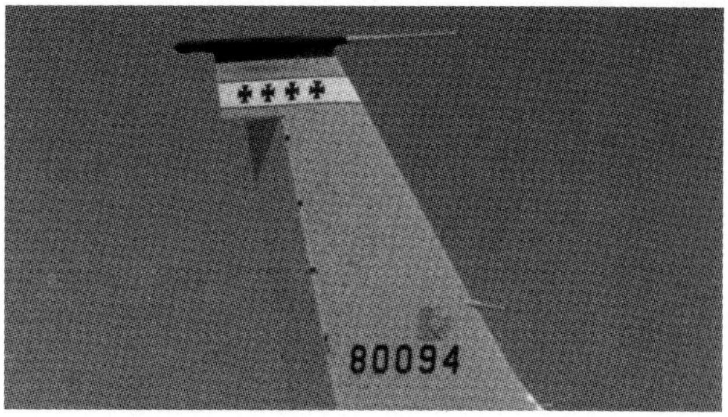

KC-135Qs from the 9th SRW at Beale AFB carry a yellow tail band with four black Maltese crosses.

In 1987, this blue, red, yellow, and orange flaming arrow fin flash was approved for KC-135As assigned to the 97th BMW at Blytheville AFB, Arkansas.

*KC-135A, 56-3653, is assigned to the 93rd BMW at Castle AFB, California. The tail marking is a small castle which was derived from the name of the base. This aircraft is named **Javelina**, and markings for it are provided in the ERTL 1/72nd scale KC-135 kit.*

*The 453rd ARefG at Fairchild AFB, Washington, uses a Seahawk as its tail marking. This is derived from Seattle's professional football team. **Fairchild** is lettered in the top band of the Seahawk marking to indicate the name of the base. Scars on the tail of this aircraft reveal that this aircraft, KC-135R, 59-1492, had served previously with the 397th BMW at Wurtsmith AFB, Michigan. The former location of that unit's Triangle K marking is clearly visible. The marking was in memory of the unit's participation in World War II when it flew B-17s out of Kimbolton, England.*

*Tail markings used by the 96th Wing at Dyess AFB, Texas include a Longhorn steer on an orange and blue tail band. Because this B-1B unit has its indigenous tankers as part of the Air Combat Command, the ACC crest appears on the fin above the **DY** tail code and the fighter-style tail number.* *(Ramsey)*

AIR NATIONAL GUARD TAIL FLASHES

Early tail markings for the New Hampshire ANG are displayed on KC-135A, 57-2608. Note the antenna above the tail number and the fin attachment for the long HF antenna wire. (157th ARefG, NHANG)

The New Hampshire ANG has changed their tail flash to this medium blue stripe with white lettering. (USAF)

This fin flash is applied to KC-135Es from the 170th ARefG. (170th AREfG, NJANG)

The 112th ARefG at Pittsburgh has this black and yellow fin flash. The ANG crest is a black stencil. This aircraft is KC-135E, 57-1479. (112th ARefG, PAANG)

The Mississippi Air National Guard Received KC-135Rs in 1992, and these tail markings were painted on the unit's aircraft. (MSANG)

MODELERS SECTION

KIT REVIEWS

1/300th SCALE KITS

Comet, Kit Numbers PL-16 and PL-26

This old model dates back to the late 1950s and was released twice using the two kit numbers listed above. It represents the Boeing Model 367-80 which is more commonly known as the Dash 80. This aircraft served as the prototype for both the 707 airliner and the -135 series of military aircraft. This kit is mentioned here only because it is the same model that was later re-released by Aurora and Addar as KC-135s. Those kits are covered next. Both of the Comet releases of this kit are now valued in the twelve to seventeen dollar range by collectors.

Aurora, Kit Number 296

With a copyright of 1963, this model originally cost thirty-nine cents. Although the box art claims that it is a model of a KC-135, it is the same as the Comet kit covered above. It even has the same engraving for the Dash 80's markings and paint scheme. There are fourteen parts and a two-piece stand which are molded in gray plastic, but no landing gear is provided.

The decals are for an NKC-135A, 55-3127, and the sheet includes a pair of insignia for the Military Airlift Command. This kit, which is now over thirty years old, is best left to the collectors who place its value in the ten to twelve dollar range.

Aurora's kit number 296 is a nothing more than a re-release of the old Comet kit in 1/300th scale. Although the box art claims that it is a KC-135, the model actually represents the Dash 80. ***(Kinzey)***

Addar, Kit Number 904

Released in 1976, this is a reissue of Comet's and Aurora's small kits which are covered above. It is packaged in a double kit along with a B-58 Hustler. Although the name on the box refers to the aircraft as a "Boeing Supertanker," the art illustrates the Dash 80. The decals are also for the Dash 80 aircraft. The parts for the Supertanker are molded in white plastic, while those for the B-58 are silver. Just because these two aircraft are packaged together, don't be misled into thinking that the two models could be used for a tiny in-flight refueling diorama. While the Dash 80 remains at about 1/300th scale, the B-58 is a much larger 1/185th scale. With the decals provided, the Dash 80 could be built into a nice but very

The same small kit that was previously released by Comet and Aurora was boxed with a B-58 Hustler and reissued by Addar in 1976. ***(Kinzey)***

small model of this famous and important aircraft. But this really isn't a kit that would be appealing to most serious scale modelers, so once again we recommend leaving it to the collectors who rate its value in the six dollar range.

1/120th SCALE KITS

Comet, Kit Number PL-80

This is a larger version of the Comet kit listed above. The model again represents the Dash 80, and it is listed here only because it was the basis for Aurora's kit number 143 which is covered next. This Comet kit is valued in the thirty to forty dollar range by collectors.

Aurora, Kit Number 143

In many respects, this kit is nothing more than an enlarged version of Aurora's kit 296 covered above. Even the painting on the box is the same as it was on the smaller kit. This is actually is a re-release of Comet's kit number PL-80, but Aurora claimed that it was a KC-135. Again, the outline represents the Dash 80 aircraft, and the markings for that aircraft are engraved into the gray plastic.

The stand for this model is clear plastic, and clear parts are provided for the windshield and windows. A crude and simplistic landing gear is also provided, but there are no wheel wells.

Aurora also released this same kit as with the numbers 143-98 and 143-100, but none of these issues offer any building potential to the serious scale modeler. Instead, their only worth is to collectors. Kit numbers 143 and 143-98 are valued in the twenty-five to thirty dollar range, but number 143-100 is listed at only twenty to twenty-five dollars.

1/144th SCALE KIT

Revell, Various Kit Numbers

The well-known Revell kit has been issued in numerous versions over the years. By changing the box art, decals, instructions, and a few pieces of plastic, Revell has released this popular model as many different 707

The Revell 1/144th scale kit was built out-of-the-box to represent KC-135A, 57-1445, from the 126th ARefW of the Ohio Air National Guard.

and -135 variants. These include, but are not limited to, a KC-135A, C-135B, E-3A AWACS, American Airlines 707-123, and even a Luftwaffe 707-307C! The latter four kits had JT3D turbofan engines.

Dating back to 1957 in its original form, this kit scales out to be about 1/143rd in span, but it is approximately 1/130th in length and height. The panel lines are generally accurate, although they are a little too heavy. The typical 1950's rivet detail must be sanded down. As a KC-135A, several corrections, improvements, and additions must be made. The boomer's pod is noticeably too large and should be replaced with a smaller scratch-built pod. However, the refueling boom is reasonably accurate. The beavertail fairings on the upper wing surfaces must be added to any Revell KC-135 kit. These fairings house a joggled wing rib under which the main gear trunnion is located.

In 1982, Revell reworked the model and released it as kit number 4523. Of all the issues, this is perhaps the best one to start with in order to model a KC-135 in 1/144th scale. It has fifty-five parts and a two-piece stand, however, the model looks better sitting on its landing gear. Some weight in the nose will eliminate the need for a pogo stick support under the tail. There is no wheel well for the nose gear, and the wells for the main gear are not deep enough. If desired, these can be added using plastic card.

Although the original issue did not have a windscreen, a clear part has been included in this newer release. Decals can be used for the few small cabin and door windows, or they can be drilled out and filled with Kristal Kleer.

J57 engines are included, replacing the TF-33 turbofans in some of the earlier issues. By substituting the JT3D fan engines from a 707 release of this kit, a KC-135E can be built. ATP makes CFM56 engines in 1/144th scale, and these could be used to represent the F108-CF-100 engines for a model of the KC-135R.

1/72nd SCALE KITS

Nova Vacuformed Kit

The first KC-135 model to be released in this popular scale, the Nova kit is vacuformed with cast metal landing gear. Fidelity of scale is quite accurate, and the kit drawing is one of the best found in any plastic model. Instructions accompanying the kit are fairly simple and straight forward. The depth of the draw on some of the parts is amazing, and the outlines are crisp and generally accurate. All of the major control surfaces and a few of the panel lines are scribed rather poorly, so some straightening of the scribe lines may be required. The airplane nose has the Boeing look, and the basic fuselage is quite good.

The wings are a five-piece design without spars. The major pieces are the left and right upper surfaces, left and right lower surfaces (extending from the inboard nacelle strut to the tip), and a lower center section.

The dihedral is molded into the two-piece horizontal stabilizer, thereby insuring proper alignment. Two different horizontal stabilizer upper surfaces are provided to account for the larger span required by turbofan-powered aircraft.

The kit comes with both J57 turbojet and TF33 turbofan nacelles. A soft metal has been used for the parts which comprise the landing gear, and this facilitates getting all ten wheels to rest in the same plane.

The Nova kit provides all of the basics and a number of extra pieces to build almost any version of the -135 family. A thimble nose is included, however, it is .25-inch too short and a bit too tapered. The boom pod comes as a separate piece for most versions of the -135. An alternative fairing has been provided for the RC-135V, RC-135U and RC-135M belly antenna. A pattern and locating information are provided for the SATCOM antenna on the kit drawing. The cheeks for the Martin-designed side-looking airborne radar (SLAR) have been included for some of the RC-135 versions. These cheeks appear to be accurate.

Decals in the Nova kit are quite complete and come on three separate sheets. These include:

1. **USAF, U. S. AIR FORCE,** national insignia, aircraft data block, and a set of numerals. Unfortunately the radio call numerals are printed in Insignia Blue and eighteen-inch gloss black numerals are difficult to find.

2. Yellow door surround stripes, a pair of SAC Milky Way bands, MATS, MAC, and SAC insignia, 93rd BMW insignia, vertical fin walkway stripes, **MAC** tail lettering, MATS/MAC nose number blocks, a series of white numerals to be applied within the blocks, and a series of white numerals for the nose.

3. **MATS** and **WEATHER** in white, **MILITARY AIRLIFT COMMAND** and **MILITARY AIR TRANSPORT SERVICE** in Insignia Blue, a pair of American flags, and a string of MATS/MAC unit identifications. These include the 1607th ATW, 1611th ATW, and 1370th PMW.

KC-135A, 62-3554, was built out-of-the-box from the ERTL kit, and it was painted and marked to represent the aircraft as it appeared when it was assigned to the 380th BMW at Plattsburg AFB, New York. It has the bicentennial markings that were used during SAC's annual bomb/nav competition "Giant Voice 1976." This aircraft, with its Thunderbird style markings on the engines, represented the overall winning unit in the competition. Decals are from Scale-Master's sheet SM-41.

AMT/ERTL, Kit Number 8848

After many years of waiting, scale modelers who attended the 1992 IPMS/USA National Convention in Seattle finally got a look at test shots for an injection-molded 1/72nd scale KC-135 kit. The first production kits were released a few months later, and the popularity of the model has been even greater than expected. The basic shape and outline is quite accurate, and the straight-forward instructions have thirteen major assembly steps.

A complete main deck floor is provided as is an aft bulkhead. This not only provides the basis for interior detailing, it also forms a structure for the alignment and rigidity of the fuselage assembly. A reasonably complete boom pod station is also included, but the instrument panel and control columns should be built from scratch and added to this pod's interior. A separate boom pod fairing is included along with a hinged visor. (Note that the visor hinge pins are very delicate and extreme care must be exercised when installing this part.) Also in the aft end is a single piece APU and a rack of oxygen bottles. The forward fuselage bulkhead is formed by the aft faces of the lavatory on the left and the electronics rack on the right. These pieces adjoin the navigator's station on the right and the electronics rack behind the boom operator's seat. Four forward flight deck seats and a pair of control columns are provided, but the folding jump seat is missing. Alternate instrument panel/aisle stands are included. One has a flat instrument panel for use with a decal, while the second is an embossed panel for the masochist to paint. The instrument panel should be vertical, however it was molded at a slight angle. The RGA console, which is located behind the pilot, is also included.

The crew entry way is also provided with an alternate position crew entry hatch. The nose wheel well dog house and gear strut are also included and must be installed when the fuselage halves are joined. Of interest is the fact that the nose wheel well doors usually hang open for maintenance, so the modeler may wish to cut them out for future separate installation. The large cargo door and a pair of hinges are provided. The Q-spring inlet is molded into the leading edge of the left vertical fin half, and care should be taken not to lose this part (it may even be better to remove the part for safe keeping for later reinstallation).

The cockpit crown is unique. Alternate clear pieces are provided, one with the P5 overhead panel molded with raised detailing and one which is smooth. However, no decal is provided for the smooth panel. A separate skull plate is provided for its location above the windscreen.

A pair of celestial observation windows are located in the cockpit overhead area. While these windows were installed in the factory, they have since been faired over. If building a later version of the KC-135, the modeler

This KC-135Q was also built out-of-the-box using the ERTL kit. It is painted in the Baby Shamu scheme, and carries a fighter-style BB tail code for "Beale Bandits." The Maltese crosses were taken from a Super Scale German F-4 decal sheet, and the 9th SRW badge is from Super Scale's sheet number 72-465. It has been toned down to represent the low visibility version of the insignia.

This ERTL model was converted to a KC-135E with the addition of JT3D (TF33-PW-102) engines that were taken from the Heller 707 kit. The horizontal stabilizers were extended by 3/8ths of an inch using sheet plastic. This aircraft is named* Hawg Wild *and is from the 161st ARefG of the Arizona Air National Guard. A refueling drogue has been attached to the refueling boom in remembrance of a TRANSPAC mission where five refuelings were required to support legless F/A-18 Hornets between Hawaii and California. The author was a guest on this particular flight.

Note the placement of the crew ladder on* Hawg Wild*. It hangs from the cockpit floor along the aft bulkhead of the entry way. It is not rotated ninety degrees against the side bulkhead nor does it rest on the ground as shown on the kit's instructions. The first rung of the ladder is a good step off of the ground. Also note the mission markers from Operation Desert Shield that are located just aft of the entrance hatch. Although the aircraft flew sorties during Operation Desert Storm, markings were not added to the aircraft to denote these missions.

should install these clear windows and paint over them with whatever finish is being applied.

The fixed portion of the refueling boom is molded in two halves, complete with the scabbed-on vortex generators. The telescoping portion of the boom is separate, as is the pair of ruddevators. A nice feature is the inclusion of the drogue adapter and separate drogue basket.

Rigidity and detail are built into the heart of the airplane. A stub of a wing front spar, with an integral forward bulkhead for the main gear wheel well bulkhead, a keel beam, and an aft bulkhead are provided for installation on the bottom of the main deck floor.

The wing assembly comes in five pieces. The lower wing center section, which has cutouts for the main gear wells, mounts onto the fuselage belly and keel assembly. Then, the left and right wing upper surfaces are installed. Care must be taken to ensure a smooth wing-to-body joint. Lastly, the lower outboard wing pieces are installed. Here it is important to ensure a good butt fit between the outboard wing panel and wing center section. Each of the four J57 turbojet assemblies comes in four pieces. Good wing alignment instructions are provided on the kit's drawing.

The horizontal stabilizer is molded in top and bottom halves, and the tiny vortex generators are molded on the lower surface. The slot for the stabilizer front spar is molded into the fuselage. Alternative fin caps are provided, one of which is the standard fin cap with the HF probe antenna. The alternative part has the new fairing for the night air refueling light molded in the aft end.

Hidden in the parts is a fuel dump tube (Part No. 118), which may be used on an RC-135A, a C-135A or a C-135B. ERTL has obviously reserved this part for use on a future kit. A pair of refueling director light assemblies is provided for installation aft of the nose gear.

The massive main gear assemblies are impressive. A one-piece main gear strut has the trunnion, drag brace, torque links, truck assembly, and axles molded together. A separate side strut is also provided. Two brake equalizer rod assemblies are included for each truck. Lastly, each main gear tire comes in two halves. The main wheel well doors are molded in the open position, but may be cut for a closed installation. However, this is not the case for the strut doors which are molded in a stacked open position.

A pogo stick is also provided for the modeler who forgot to add weight in the nose or for the one who

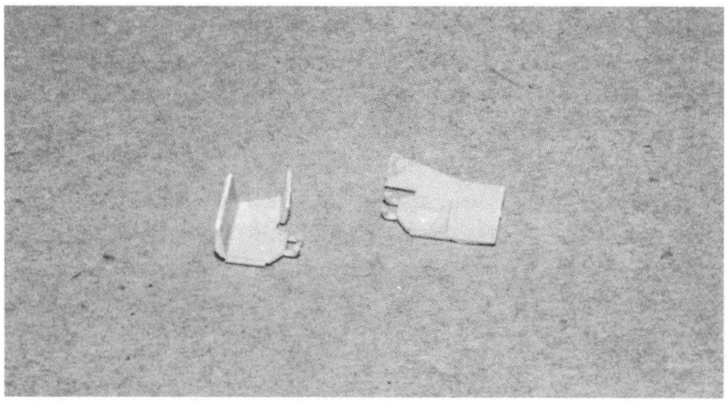

ERTL's instruction fail to show the proper positions for the main landing gear strut doors. The correct positions are required for both scale accuracy and fit around the gear when it is extended. At left is a door that has been bent to the correct angles, while at right is the flat door as it comes in the kit.

***Eeyore** is a KC-135A from the 92nd BMW at Fairchild AFB, Washington. This model was built out-of-the-box from the ERTL kit. With its overall gray paint scheme, it is a clean and colorful model with nice eye appeal.*

wishes to add a little more detail. Incidentally, the pogo stick is a piece of ground support equipment and therefore is painted yellow, except for the ones used with EC/RC-135s. The ones used with these variants are red because of the proximity to the radiation hazard area.

Step 13 shows the crew ladder facing inboard. It would be difficult at best to get into the aircraft with the ladder in this position. The ladder should be attached to a fitting on the aft wall of the crew entry way. When installed, the lower rung of the ladder is a good step above the ground.

The kit comes with alternative markings as described on AMT/ERTL Form Number 85-0997.

1. The overall Aircraft Gray (FS 16473) aircraft is KC-135A, 57-1423. Markings are for the 92nd ARefS, 92nd BMW, which is based at Fairchild AFB, Washington. The aircraft was named **Eeyore** for the character in A. A. Milne's Winnie the Pooh. In addition to the standard SAC markings, the decal sheet includes the Seahawk emblem for the sides of the vertical fin and the name plate which goes next to the crew entry hatch. The radome should be painted gloss black (FS 17038), while the anti-glare panel is flat black (FS 37038). Decal No. 9 is the data block for this aircraft. It should be noted that the SAC and 92nd BMW insignia are located too high on the Milky Way bands and must be lowered accordingly. Instructions for application of the two-inch wide wing and horizontal stabilizer walkways are shown on the instruction sheet.

2. The Baby Shamu scheme is for KC-135A, 56-3653, which has markings for the 93rd ARefS, 93rd BMW, at Castle AFB, California. The lower surface of the aircraft should be painted gloss Aircraft Gray (FS 16743), while the upper surface should be painted flat Dark Gray (FS 36081). If aged, some dark green should be added. The radome should be gloss black (FS 17038), while the anti-glare panel is flat black (FS 37038). Standard low visibility aircraft markings are provided. A pair of Castles, which are representative of Castle AFB, are included for each side of the vertical fin. Decal No. 35 is the data block for this aircraft. Decal No. 33 has the crew chief's name and the aircraft name **Javelina**. Note that the instructions do not point out the fact that Decal No. 20, the refueling alignment stripe, must also be applied to the belly of this aircraft. Also, Decal No. 10 must be modified to provide the walkway marking on the vertical fin of any KC-135 in the Baby Shamu scheme. In addition, the walkway markings for the upper surfaces of the wing and horizontal stabilizer must also be added.

ACCESSORIES & CONVERSION KITS

Note: All of the accessories and conversion kits listed below are in 1/72nd scale, and they are intended for use with the ERTL kit.

True Details, Kit Number 45007

This cast resin set provides eight main wheel and two nose wheel assemblies. The detailing is extremely crisp and the brake assemblies in the inner surfaces are nicely represented. In addition, the wheel cooling holes and tire treads are molded into the parts. However, while all of the tires are flattened to indicate a load on them, they are also bulged. These bulges should not be present.

The kit costs $4.98 plus postage and is available through Lencraft, P.O.Box 770, Springville, California 93265. It may also be obtained from Squadron Mail Order, 1115 Crowley Drive, Carrollton, Texas 75011.

True Details, Kit Number 2505

This photo-etched brass detail set includes a variety of UHF, VHF and TACAN antennas, angle of attack vanes, and pitot probes. The modeler must consult photographs for specific installation locations on the different KC-135 variants.

At $3.50, this detail set is also available through Lencraft or Squadron Mail Order.

True Details, Kit Number 45001

This photo-etched brass detail set contains ten instrument panel pieces, four sets of seat belts, a glare shield, two circuit breaker panels, a nose gear steering wheel, entry door crank and chinning bar, entry chute floor grill, a pair of windshield wipers, two pitot masts, a water drain mast, and doors for the main gear struts.

The retail price is $6.98, and this set is also available through Lencraft or Squadron Mail Order.

True Details, Kit Number 45012

This resin kit provides a complete galley with trash bin, oven, and wheel chocks. Also included are warning horns, pressure release port, cargo door controls, air conditioning duct and controls, and an optional book case.

At $4.50, this detail set is also available through Lencraft or Squadron Mail Order.

Airwaves Accessories, Number 72100

This photo-etched brass detail set provides three

fretts which contain a plethora of parts for the basic airplane as well as antennas for the KC-135A, KC-135R, EC-135C, RC-135V and RC-135S. Antennas for the latter are only those visible for the left (non-working side) of the aircraft.

The basic parts include seat belts, instrument panels, control column level, floor grill, a dubious two-piece crew access ladder, cargo door barrier net, a frame for the wing leading edge landing light, compressor blades (erroneously identified as turbine blades), windshield wipers, and cargo door actuators.

A careful study of the parts prior to installation is strongly recommended. At $17.95, this detail set is also available through Lencraft or Squadron Mail Order.

<u>Maintrack Conversion Kit Number 72:29</u>

Four TF33 turbofan nacelles, two humped and two unhumped engine struts, and a pair of larger horizontal stabilizers are included in this kit. The stabilizers are about 3/8ths of an inch too large in span and chord. Maintrack is correcting these parts in future releases, and they will replace the incorrect stabilizers in the earlier kits. This kit is available through Lencraft at a price of $39.99 plus postage.

<u>Maintrack Accessory Pack 26</u>

A single humped outboard strut for a TF33 engine is provided in this accessory pack. It is available for $5.00 and postage from Lencraft.

<u>Maintrack Accessory Pack 27</u>

This pack provides a single unhumped outboard strut for a TF33 engine. It can also be ordered from Lencraft for $5.00 plus postage.

<u>Maintrack Accessory Pack 28</u>

A replacement fin cap for an early short-tailed KC-135A is included in this pack. It sells for $5.00 and can be ordered from Lencraft.

<u>ATGR Conversion Kit</u>

This kit allows conversion of the ERTL KC-135A kit to a KC-135R. It contains four resin nacelle/strut assemblies, each molded in two parts. These are to be mated midway along their length. The conical spinner and inlet fan are deeply recessed into the forward half of the engine. A series of stepped exhaust cones are molded into the aft half.

Instructions are provided for extending the horizontal stabilizers and fabricating the APU inlet and exhaust doors. A color diagram is included for the refueling director lights and the strobe beacon. In addition, instructions explain how to cut the inner main gear doors so that they can be displayed in the open position.

This kit retails for $21.50, and it may be purchased directly from ATGR Enterprises, Inc., 407 Van Roen Road, Utica, New York 13502.

<u>DB Conversions, Kit Number 2028</u>

This kit is a set of resin wheel assemblies with flattened bottoms.

<u>DB Conversions, Kit Number 2029</u>

Four engine inlets, each with an integral set of inlet stator vanes and a nose dome, are provided in this kit. The oil cooler inlet has a crisper opening than the ones in the ERTL kit. The inlet lining is the more accurate, because it is completely circular. By comparison, the kit inlet drops below the inlet stator vanes.

USING PARTS FROM OTHER KITS FOR CONVERSIONS

<u>Heller Boeing 707 or AWACS Kits</u>

This is the expensive way to go, but a modeler may wish to use the JT3D turbofan engines and struts from one of these kits to build a KC-135E. In actuality, it would be best to use the AWACS kit and replace that kit's engines with DB Conversions' TF33s which are accurate. Only a set of struts would then need to be scratch-built for the AWACS.

<u>Airfix E-3D AWACS Kit</u>

While more expensive than the Heller 707 and AWACS kits, the Airfix kit comes with both JT3D and CFM56 engines.

The ERTL KC-135A was converted to a KC-135R using DB Conversion's CFM56 engines. This conversion kit was originally made for the Heller 707 kit, and some scale differences may account for difficulties when hanging these engines on the ERTL KC-135 model. Extreme care must be taken in mounting the engines to ensure a good tight fit to the wings and to prevent them from hanging too low. Some reshaping to the tops of the struts may be required. The kit's horizontal stabilizers were extended by 3/8ths of an inch in span using sheet plastic. The tail markings were hand painted, and the 301st ARefW insignia came from a Hasegawa B-47 decal sheet that was toned down for low visibility.

DECALS

Only two decal sheets have been issued specifically for KC-135 models. They are both in 1/72nd scale.

Scale-Master, Sheet Number SM 41

Entitled *Bicentennial KC-135As*, this sheet provides markings for three aircraft.

KC-135A, 62-3554, **Spirit of North Country**, 380th BMW, Plattsburgh AFB, New York

KC-135A, 57-2591, **Kitty Hawk**, 68th BMW, Seymour Johnson AFB, North Carolina

KC-135A, 63-7981, **Center City USA**, 384th ARefW, McConnell AFB, Kansas

Three Guys Replicas, Sheet Number 3G-001

Markings for four KC-135As and one KC-135Q are included on this sheet.

KC-135A, 57-1460, 11th ARefS, with shark's teeth, circa 1972

KC-135A, 61-0300, 22nd ARefW, flagship for the 15th Air Force, circa 1988

KC-135A, 62-3509, **Spirit of the Seacoast**, circa 1990

KC-135A, 59-1481 (N903NA), NASA, circa 1986. A typographical error on the instruction sheet indicates that the serial number is 59-1591. This aircraft was named **Weightless Wonder IV** and was used for astronaut training. This boomless aircraft may be built straight out of the box, because the fuel dump tube is provided with the ERTL kit.

KC-135Q, 59-1523, 9th SRW, circa 1988

Alternative decals may be found on several after-market decal sheets and the decal sheet in the Hasegawa B-47E kit as follows:

7th BMW	Super Scale (B-52 sheet), 72-152
9th SRW	Super Scale, 72-465
19th BMW	Almark, S18
22nd ARefW	1992 IPMS National Convention Decal
28th BMW	Super Scale, 72-577 & 72-581
42nd BMW	Almark, S18
43rd SW	Almark, S18
92nd BMW	Almark, S18
96th BMW	Super Scale, 72-561, 72-577, & 72-581
98th BMW	Super Scale, 72-561
301st BMW	Hasegawa B-47E kit
310th BMW	Super Scale, 72-561
320th BMW	Almark, S18
AFFTC	Almark, S17
ARDC	Almark, S17

Additionally, **UNITED STATES AIR FORCE** can be found on A.I.R. sheet number 9. **U. S. AIR FORCE** is on A.I.R. sheet number 9 and Super Scale sheet numbers 72-137 and 72-188. **MATS, MAC, AFSC, AFLC, PACAF, USAFE**, and **ANG** markings are on Super Scale sheets 72-187 and 72-190. Black walkway stripes can be made using decal stripes on Scale-Master sheet number SS-2. Use the stripes identified by the letter G.

REFERENCES

Bell, Dana — USAF Colors and Markings in the 1990s, Greenhill Books, London, 1992

Bowers, Peter M. — Boeing Aircraft Since 1916, Putnam, London, 1989

Coyne, James P. — Airpower in the Gulf, Air Force Association, Arlington, Virginia, 1992

Flintham, Victor — Air Wars and Aircraft, a Detailed Record of Air Combat, 1945 to the Present, Facts on File, New York, 1990

Kinzey, Bert — The Fury of Desert Storm - The Air Campaign, Kalmbach, Waukesha, Wisconsin, 1991

Lloyd, Alwyn T. — Boeing 707 and AWACS in Detail & Scale, Kalmbach, Waukesha, Wisconsin, 1987

Micheletti, Eric — Operation Daguet - French Air Force in the Gulf War, Concord Publications, No.1022, Hong Kong, 1991

Morse, Stan — Gulf Air War Debrief, World Airpower Journal, Airospace Publiahing, Ltd., London, 1991

Reed, C. M. — KC-135 Stratotanker in Action, No 118, Squadron/Signal, Carollton, Texas, 1991

Lloyd, Alwyn T. — *Versatility Unlimited - The Boeing KC-135 Story,* AIR International, London, Part 1, November 1980; Part 2, December 1980

Lloyd, Alwyn T. — *30 Years Young - The KC-135 Celebrates 30 Years of Flight,* AAHS Journal, Fall 1987